云原生
Kubernetes
自动化运维实践

亨鹏举 著

清华大学出版社
北京

内 容 简 介

本书以一名大型企业集群运维工程师的实战经验为基础,全面系统地阐述Kubernetes(K8s)在自动化运维领域的技术应用。本书共16章,内容由浅入深,逐步揭示K8s的原理及实际操作技巧。第1章引领读者踏入Kubernetes的世界,详细介绍其起源、核心组件的概念以及集群安装方法。第2～4章深入剖析Pod控制器、Label标签、容器钩子、探针、Service服务发现与负载均衡机制。第5～7章则探讨Ingress-Nginx服务网关的应用、存储卷管理、配置和密钥管理的高级功能。第8～10章聚焦于鉴权机制、容器运行时的选择与配置,以及GitLab企业级代码仓库的部署和管理。第11～13章涵盖Jenkins持续集成交付工具、ArgoCD声明式持续交付,以及云原生负载均衡MetalLB的应用。第14章和第15章分别介绍云原生日志与监控集成架构,以及Istio微服务时代的服务网格领航者。最后,第16章通过一系列实战案例,展示在K8s环境中安装并实验多种服务的详细过程。

本书不仅整合了多种自动化运维工具,还提供了丰富的运维案例,无论是初学者还是有一定经验的运维工程师,都能从中获得宝贵的知识和实践经验,提升自身的技术水平。

本书封面贴有清华大学出版社防伪标签,无标签者不得销售。
版权所有,侵权必究。举报: 010-62782989, beiqinquan@tup.tsinghua.edu.cn。

图书在版编目(CIP)数据

云原生Kubernetes自动化运维实践 / 高鹏举著. -- 北京:清华大学出版社, 2025.1. -- ISBN 978-7-302-67934-9
Ⅰ. TP316.85
中国国家版本馆CIP数据核字第20257QE007号

责任编辑:王金柱
封面设计:王　翔
责任校对:闫秀华
责任印制:刘　菲

出版发行:清华大学出版社
网　　址:https://www.tup.com.cn,https://www.wqxuetang.com
地　　址:北京清华大学学研大厦A座
邮　　编:100084
社 总 机:010-83470000
邮　　购:010-62786544
投稿与读者服务:010-62776969,c-service@tup.tsinghua.edu.cn
质量反馈:010-62772015,zhiliang@tup.tsinghua.edu.cn

印 装 者:三河市少明印务有限公司
经　　销:全国新华书店
开　　本:185mm×235mm　　印　张:22　　字　数:528千字
版　　次:2025年3月第1版　　印　次:2025年3月第1次印刷
定　　价:99.00元

产品编号:107451-01

前　　言

——高效运维，轻松适应云原生时代

在这个云原生与DevOps风起云涌的时代，技术的浪潮不断推动着软件开发的边界，让我们见证了前所未有的变革与创新。作为一名深耕运维开发领域的工程师，笔者深知在这片浩瀚的技术海洋中，每一步探索都充满了挑战与机遇。因此，笔者怀揣着对技术的热爱与敬畏，决定将这段旅程中的所见所感，以及实战经验，汇聚成这本《云原生Kubernetes自动化运维实践》。

本书精心编排了16个章节，从云原生技术的基石——Kubernetes（K8s）出发，逐步深入Containerd容器化技术、版本控制系统GitLab、持续集成/持续部署（CI/CD）工具Jenkins，再到监控与日志分析领域的强大组合Loki+Grafana+Prometheus，每一个章节都是对运维开发领域关键技术的深刻剖析与实践总结组成部分。

我们不会止步于此。随着云原生架构的日益成熟，负载均衡与服务发现成为不可忽视的议题，MetalLB以其独特的魅力为我们提供了云外Kubernetes集群的负载均衡解决方案。而Argo CD作为GitOps的杰出代表，将Git仓库作为基础设施和应用声明的唯一事实来源，引领了配置管理的新风尚。最后，当我们踏入服务网格的殿堂，Istio以其强大的流量管理、安全通信和策略控制功能，为我们构建分布式系统提供了前所未有的灵活性和可靠性。

撰写此书的过程中，笔者力求做到理论与实践相结合，既有深入浅出的技术讲解，也有丰富的实战案例。无论你是刚刚踏入运维开发领域的新手，还是有一定经验的运维人员，都能从这本书中找到属于自己的收获。

本书结构

本书共分为16章，循序渐进地为读者呈现从基础原理到实战应用的全面解析。

第1章走进Kubernetes的世界，介绍K8s的由来，以及K8s的组件概念、集群安装。

第2章玩转Pod控制器，介绍Kubernetes的各类Pod控制器的概念以及使用方法。

第3章Label、容器钩子、探针与HPA，介绍Label、钩子、探针、自动扩容类型和概念，通过各类实验带你了解在各个场景下如何选择合适的控制器来运行服务。

第4章Service服务发现与负载均衡，介绍Service如何作为Pods的逻辑集合的访问入口，深入探讨Service的几种类型。

第5章Ingress-Nginx服务网关，介绍安装和配置Ingress-Nginx，并演示如何通过定义Ingress资源对象来创建和管理路由规则。此外，我们还将探讨一些高级配置和最佳实践，帮助读者优化Ingress-Nginx的性能和安全性。

第6章Kubernetes存储与持久化，介绍存储卷的基本概念与作用，掌握如何安装并配置和使用StoreClass来实现动态申请存储。

第7章Kubernetes配置和密钥管理，介绍ConfigMap和Secret的基本概念、用途、创建方式以及使用场景。

第8章探索Kubernetes鉴权机制，介绍Kubernetes的鉴权体系，包括认证（Authentication）和授权（Authorization）两大核心环节，以及它们如何协同工作以确保只有经过授权的用户或系统才能够访问集群资源。

第9章探索云原生时代的容器运行时，介绍Containerd的基本概念，如何安装和配置Containerd，以及安装并使用nerdctl工具来对Containerd进行管理和安装，并使用BuildKit工具来构建镜像。

第10章GitLab企业级代码仓库，介绍如何在K8s集群上部署GitLab这一流行的开源Git仓库管理工具，以及如何使用Git工具通过命令行操作GitLab来提交克隆和提交代码。

第11章Jenkins持续集成交付工具，介绍如何在K8s环境中安装Jenkins，Jenkins与Kubernetes集群对接以实现动态slave（工作节点）的创建与管理。

第12章ArgoCD声明式持续交付，介绍如何在K8s集群中安装并配置ArgoCD，实现应用部署的自动化与版本控制。

第13章云原生负载均衡之MetalLB，介绍如何在K8s集群中安装MetalLB以及使用LoadBalancer类型Service的Kubernetes应用，并验证MetalLB是否正确地为该Service分配了IP地址。

第14章探索云原生日志与监控集成架构，介绍如何在K8s集群中利用Helm包管理工具来安装和配置Prometheus、Grafana与Loki日志系统以及配置监控告警。

第15章Istio微服务时代的服务网格领航者，介绍如何在K8s集群中安装Istio服务，展示了如何使用Istio的流量路由功能，包括基于权重的路由分发、故障注入和蓝绿部署等。

第16章深入探索K8s服务部署实战，介绍在K8s环境中安装并实验多种服务的详细过程，展示了如何编写相应的K8s资源清单文件，如Deployments、Services、ConfigMaps、Secrets等。

资源下载

本书提供配套资源，读者用微信扫描下面的二维码即可获取。

如果读者在学习本书的过程中遇到问题，可以发送邮件至booksaga@126.com，邮件主题为"云原生Kubernetes自动化运维实践"。

读者对象

本书适合Kubernetes初学者、运维人员、开发人员，以及对运维技术感兴趣的各类读者阅读。

最后，感谢所有在笔者学习和写作过程中给予帮助和支持的人，是你们的鼓励与指导，让我有勇气将这份热爱化作文字，与读者分享。

若你在阅读本书的过程中有任何问题或者建议，可以关注"院长技术"公众号并加入微信群与笔者交流。

让我们携手并进，在运维开发的征途中，不断探索、学习、成长，共同迎接更加辉煌的未来！

高鹏举
2025年1月

目 录

第 1 章 走进 Kubernetes 的世界 1
1.1 为什么使用 Kubernetes 1
1.2 Kubernetes 节点组件 2
1.2.1 Master 节点运行的组件 2
1.2.2 Node 节点运行的组件 2
1.3 Kubernetes 集群的搭建与配置 3
1.3.1 Kubeadm 工具介绍 3
1.3.2 基础环境配置 4
1.3.3 升级内核版本 10
1.3.4 三大组件的安装 12
1.3.5 集群镜像处理 13
1.3.6 集群初始化 14
1.3.7 Node 节点加入集群 14
1.3.8 Calico 网络插件安装 16
1.3.9 Metrics-Server 服务的安装 17
1.3.10 Kuboard 管理平台安装 20
1.4 本章小结 22

第 2 章 Pod 控制器 23
2.1 Pod、Kubectl 与 YAML 23
2.1.1 Pod 及其操作 24
2.1.2 Kubectl 命令行工具 25
2.1.3 YAML 文件 27
2.2 Replication 28
2.3 ReplicaSet 31
2.4 Deploymant 36
2.4.1 Deployment 概述 36
2.4.2 Deployment 创建与访问 36
2.4.3 滚动更新策略实例 38
2.4.4 更新/回滚/暂停/恢复 41
2.4.5 扩容的 3 种方式 45

2.5 StatefulSet ·· 46
2.5.1 StatefulSet 概述 ·· 46
2.5.2 StatefulSet 创建服务 ·· 47
2.5.3 Ping 域名实验 ··· 49
2.5.4 解析域名实验 ··· 50
2.5.5 创建 StatefulSet 服务自动申请 PV 实验 ·· 51
2.6 DaemonSet ··· 55
2.6.1 DaemonSet 概述 ··· 55
2.6.2 DaemonSet 实例 ··· 56
2.7 CronJob ·· 57
2.7.1 CronJob 概述 ·· 57
2.7.2 CronJob 资源清单详解 ·· 58
2.7.3 CronJob 实验 ·· 59
2.7.4 CronJob 实战备份 MySQL 数据库 ··· 61
2.8 Job ·· 64
2.8.1 Job 概述 ·· 64
2.8.2 Job 实验 ·· 64
2.9 本章小结 ··· 66

第 3 章 Label、容器钩子、探针 ·· 67
3.1 Label 标签 ··· 67
3.1.1 Label 概述 ··· 67
3.1.2 Label 实验 ··· 68
3.2 InitC ·· 70
3.2.1 InitC 概述 ·· 71
3.2.2 InitC 实验 ·· 71
3.2.3 部署 Elasticsearch 服务时配置 InitC ·· 73
3.3 容器钩子 ··· 75
3.3.1 容器钩子概述 ··· 75
3.3.2 容器钩子实验 ··· 76
3.4 探针 ··· 78
3.4.1 探针概述 ··· 78
3.4.2 StartUp Probe 启动探针实验 ··· 79
3.4.3 Readiness Probe 就绪探针实验 ·· 80
3.4.4 LivenessProbe 存活探针实验 ·· 82
3.5 本章小结 ··· 83

第 4 章 Service 服务发现与负载均衡 ... 84

4.1 Service 原理 ... 84
4.2 ClusterIP ... 85
4.2.1 ClusterIP 概述 ... 85
4.2.2 ClusterIP 实验 ... 86
4.3 NodePort ... 88
4.3.1 NodePort 概述 ... 88
4.3.2 NodePort 实验 ... 89
4.4 Headless Service ... 91
4.4.1 Headless Service 概述 ... 91
4.4.2 Headless Service 实验 ... 92
4.5 ExternalName ... 94
4.5.1 ExternalName 概述 ... 95
4.5.2 ExternalName 实验 ... 95
4.6 LoadBalancer ... 98
4.6.1 LoadBalancer 概述 ... 98
4.6.2 如何指定 LoadBalancer 类型的服务 IP ... 98
4.7 Service 端口范围及解除限制 ... 99
4.7.1 Service 端口范围概述 ... 99
4.7.2 Service 端口范围解除限制 ... 99
4.8 使用 Service 代理 K8s 外部应用 ... 100
4.8.1 使用 Service 代理 K8s 外部应用概述 ... 101
4.8.2 使用 Service 代理 K8s 外部应用实验 ... 101
4.9 本章小结 ... 103

第 5 章 Ingress-Nginx 服务网关 ... 104

5.1 Ingress-Nginx 概述 ... 104
5.2 Ingress-Nginx 安装 ... 104
5.3 Annotations 注解 ... 109
5.3.1 流量复制 ... 109
5.3.2 IP 白名单 ... 110
5.3.3 IP 黑名单 ... 111
5.3.4 域名转发 ... 111
5.3.5 返回字符串 ... 112
5.3.6 文件上传大小 ... 113
5.3.7 域名 HTTPS 访问 ... 113

5.3.8 对接外部的认证服务 ··· 115
5.3.9 配置默认页面 ··· 116
5.3.10 Nginx 如何获取客户端真实 IP ··· 117
5.3.11 重定向 ·· 117
5.3.12 重写 ·· 118
5.3.13 多域名指向同一个后端服务 ·· 118
5.4 本章小结 ··· 119

第 6 章 Kubernetes 存储与持久化 ·· 120

6.1 Kubernetes 存储类概述 ·· 120
6.2 Kubernetes 持久卷声明 ·· 121
6.3 持久卷的生命周期 ·· 121
6.4 动态申请持久卷实验 ··· 122
    6.4.1 NFS 共享存储搭建 ·· 122
    6.4.2 nfs-client-provisioner 存储类搭建 ··· 124
    6.4.3 服务使用存储类动态申请资源实验 ·· 129
6.5 PV/PVC 详解 ·· 131
    6.5.1 PV/PVC 概述 ··· 131
    6.5.2 PVC 的创建流程 ·· 131
    6.5.3 PV 访问模式 ··· 132
    6.5.4 PV 回收策略 ··· 132
    6.5.5 PV/PVC 卷状态 ·· 133
6.6 Deployment 直连 NFS 存储 ··· 134
6.7 本章小结 ··· 135

第 7 章 ConfigMap 配置和 Secret 密钥管理 ··· 136

7.1 ConfigMap：非敏感配置信息的集中管理 ·· 136
    7.1.1 ConfigMap 概述 ·· 136
    7.1.2 使用目录方式创建 ConfigMap ·· 137
    7.1.3 使用文件方式创建 ConfigMap ·· 139
    7.1.4 使用字面值方式创建 ConfigMap ··· 139
    7.1.5 设置 ConfigMap 不允许更改 ··· 140
    7.1.6 通过 envfrom 方式指定 ConfigMap ··· 140
    7.1.7 通过 valueFrom 方式指定 ConfigMap ·· 142
    7.1.8 Nginx 通过 ConfigMap 管理配置文件 ·· 143
7.2 Secret：敏感信息的安全存储与访问 ·· 147
    7.2.1 Secret 概述 ··· 147

7.2.2	使用文件方式创建 Secret	147
7.2.3	使用 YAML 方式创建 Secret	149
7.2.4	Secret 权限解析	149
7.2.5	使用 Docker 的 config.json 方式创建 Secret	149
7.2.6	使用 Kubectl 创建 Docker Registry 认证的 Secret	150
7.3	本章小结	151

第 8 章 Kubernetes 鉴权机制 ... 152

- 8.1 Kubernetes 鉴权机制概述 ... 152
- 8.2 鉴权机制的工作流程 ... 152
- 8.3 角色/角色绑定概述 ... 153
- 8.4 用户鉴权实战 ... 154
- 8.5 本章小结 ... 160

第 9 章 容器运行时 Containerd ... 161

- 9.1 Containerd 概述 ... 161
- 9.2 安装与配置 Containerd ... 162
  - 9.2.1 安装 Containerd ... 162
  - 9.2.2 配置 Containerd 阿里云镜像加速器 ... 164
  - 9.2.3 配置 Containerd 使用自建镜像仓库 ... 164
- 9.3 使用 nerdctl 管理 Containerd ... 165
  - 9.3.1 安装 nerdctl ... 165
  - 9.3.2 nerdctl 常用命令示例 ... 166
- 9.4 使用 nerdctl 构建镜像 ... 167
  - 9.4.1 安装 BuildKit 和 cni-plugins ... 168
  - 9.4.2 构建镜像 ... 169
- 9.5 本章小结 ... 170

第 10 章 GitLab 企业级代码仓库 ... 171

- 10.1 GitLab 目录结构 ... 171
- 10.2 部署 GitLab ... 172
- 10.3 GitLab 的配置与使用 ... 183
  - 10.3.1 基础设置 ... 183
  - 10.3.2 创建项目 ... 184
  - 10.3.3 修改克隆地址 ... 185
  - 10.3.4 拉取/提交代码 ... 188
- 10.4 本章小结 ... 190

第 11 章 Jenkins 持续集成交付工具 191
11.1 Jenkins 概述 191
11.2 Kubernetes 集群部署 Jenkins 192
11.3 Jenkins 对接 K8s 实现动态 Slave 199
11.3.1 基础设置并对接 K8s 199
11.3.2 自由风格项目实现动态 Slave 203
11.3.3 Pipeline 流水线项目实现动态 Slave 205
11.4 本章小结 208

第 12 章 ArgoCD 声明式持续交付 209
12.1 ArgoCD 概述 209
12.2 Kubernetes 部署 ArgoCD 210
12.3 ArgoCD 的配置及使用 215
12.3.1 ArgoCD 连接 Kubernetes 215
12.3.2 使用 ArgoCD CLI 集成 GitLab 并创建 App 215
12.4 本章小结 221

第 13 章 云原生负载均衡之 MetalLB 222
13.1 自建 LoadBalancer 种类 222
13.2 MetalLB 的核心概念与架构 223
13.2.1 MetalLB 的核心概念 224
13.2.2 MetalLB 架构 224
13.3 Kubernetes 部署 MetalLB 225
13.3.1 检查是否开启 IPVS 功能 225
13.3.2 配置并创建 MetalLB 服务 226
13.3.3 创建 LoadBalancer 类型的服务 228
13.3.4 使用 MetalLB 进行服务的外部访问 229
13.4 本章小结 230

第 14 章 Helm 与 Loki-Stack 搭建日志监控系统 231
14.1 Helm 包管理与部署 231
14.1.1 Helm 概述 231
14.1.2 CentOS 7 系统安装 Helm3 232
14.2 Loki-Stack 部署与实践 233
14.2.1 Loki 与 Loki-Stack 概述 233
14.2.2 Helm3 部署 Loki-Stack 234
14.2.3 外部访问 Grafana 237

		14.2.4	日志监控查询	238

- 14.2.4 日志监控查询 ··········· 238
- 14.2.5 导入仪表盘面板 ········· 239
- 14.2.6 监控告警 ··············· 242
- 14.3 本章小结 ····················· 246

第 15 章 Istio 微服务时代的服务网格领航者 247

- 15.1 Istio 概述 ···················· 247
- 15.2 Istio 核心组件 ················ 248
  - 15.2.1 Istio-Pilot ············· 248
  - 15.2.2 Istio-Telemetry ········ 248
  - 15.2.3 Istio-Policy ············ 249
  - 15.2.4 Istio-Citadel ··········· 249
  - 15.2.5 Istio-Sidecar-Injector ··· 250
  - 15.2.6 Istio-Proxy ············ 250
  - 15.2.7 Istio-Ingress-Gateway ··· 251
  - 15.2.8 Istio-Envoy ············ 251
- 15.3 部署 Istio ···················· 251
  - 15.3.1 Istioctl 的安装 ········· 251
  - 15.3.2 Istioctl 安装 Istio ······ 253
- 15.4 Sidecar 边车容器注入 ········· 256
  - 15.4.1 Sidecar 手动注入 ······· 256
  - 15.4.2 Sidecar 自动注入 ······· 259
  - 15.4.3 Sidecar 取消自动注入 ··· 260
- 15.5 4 种配置资源概念详解 ········· 261
  - 15.5.1 VirtualService ·········· 261
  - 15.5.2 DestinationRule ········ 261
  - 15.5.3 ServiceEntry ··········· 262
  - 15.5.4 Gateway ··············· 262
- 15.6 VirtualService 关键字配置示例 ··· 262
  - 15.6.1 使用 weight 关键字拆分流量 ··· 262
  - 15.6.2 使用 timeout 关键字设置请求超时时间 ··· 263
  - 15.6.3 使用 retries 关键字设置重试 ··· 264
  - 15.6.4 使用 fault 关键字设置故障注入 ··· 264
  - 15.6.5 VirtualService 资源清单详解 ··· 265
- 15.7 Istio 流量治理 ················ 265
  - 15.7.1 请求头 httpHeaderName ··· 266
  - 15.7.2 HTTP 流量镜像 ········· 269

- 15.7.3 重写 ·· 274
- 15.7.4 重定向 ·· 277
- 15.7.5 流量权重—蓝绿与金丝雀发布 ····· 282
- 15.7.6 超时 ·· 287
- 15.7.7 重试 ·· 293
- 15.7.8 断路器/熔断 ·································· 303
- 15.7.9 故障注入 ······································ 313
- 15.8 本章小结 ··· 316

第 16 章 Kubernetes 服务部署实战 ········· 317

- 16.1 K8s 部署 MinIO 开源对象存储 ············· 317
- 16.2 K8s 部署 Metabase 数据库连接工具 ····· 321
- 16.3 K8s 部署 phpMyAdmin 数据库连接工具 ····· 326
- 16.4 K8s 部署 Nacos 配置中心 ······················ 330
- 16.5 本章小结 ··· 338

第 1 章

走进Kubernetes的世界

本章将带领读者进入Kubernetes的世界。我们将重点讲解为什么使用Kubernetes及其组件概念，带领读者了解如何快速搭建Kubernetes集群，并介绍Metrics服务的安装和使用等内容。

1.1 为什么使用Kubernetes

Kubernetes这个单词起源于希腊语，意为"舵手"或"领航员"，它是"管理者"和"控制论"的根源。Google于2014年将Brog系统开源为Kubernetes，简称为K8s，K8s的含义是用单个字符8来代替ubernete中的8个字符。

Kubernetes是一个可移植、可扩展的开源平台，它使用声明式配置，并依据配置信息自动管理容器化应用程序。在所有容器编排工具中（如Docker Swarm、Mesos等），Kubernetes拥有更大的生态系统和更快的增长速度，提供了更多的支持、服务和工具供用户选择。

作为一个开源的容器编排平台，目前Kubernetes已广泛应用于生产环境中，其主要解决的问题如下。

- 服务发现和负载均衡：Kubernetes可以使用标签或DNS名称来暴露容器，并且在流量较大时进行负载均衡和服务发现。
- 存储编排：Kubernetes允许自动挂载所选的存储系统，比如本地存储、公共云提供商的存储等。
- 自我修复：Kubernetes能够重新启动失败的容器、进行替换并将它们重新调度到其他节点上。
- 密钥与配置管理：Kubernetes存储和管理敏感信息，如密码、OAuth令牌和SSH密钥，允许在不重建镜像的情况下更新和部署密钥和应用配置。
- 横向扩展：使用简单的命令、用户界面或自动化构建系统，可以快速增加或减少副本数。

- 集群管理：Kubernetes能够在多个主机上运行容器化应用，提供服务发现和负载均衡，以及跨区域的故障转移。

Kubernetes通过提供这些功能，极大地简化了容器化应用的部署和管理，使得开发者和运维人员能够更加专注于应用的开发和维护，而不是底层基础设施的细节。

1.2 Kubernetes节点组件

在Kubernetes中，节点是集群中的物理机或虚拟机，它们负责运行容器化的应用程序。每个节点都由一组组件组成，这些组件负责管理节点上容器的生命周期和网络通信。本节将介绍Kubernetes中的节点组件，包括Master节点组件和Node节点组件。通过了解这些组件的作用和功能，读者可以更好地理解Kubernetes的工作原理和架构设计。

1.2.1 Master节点运行的组件

Kubernetes中的Master指的是集群控制节点，在每个Kubernetes集群中都需要有一个Master来负责整个集群的管理和控制，基本上Kubernetes的所有控制命令都发给它，它负责具体的执行过程，我们后面执行的所有命令基本都是在Master上运行的。Master通常会占据一个独立的服务器（高可用部署建议用3台服务器），主要是因为它太重要了，是整个集群的"首脑"，如果它宕机或者不可用，那么对集群内容器应用的管理都将失效。

下面详细介绍在Master上运行的关键组件。

- Kubernetes API Server（kube-apiserver）：提供了HTTP REST接口的关键服务进程，是Kubernetes中所有资源的增、删、改、查等操作的唯一入口，也是集群控制的入口进程。
- Kubernetes Controller Manager（kube-controller-manager）：Kubernetes中所有资源对象的自动化控制中心，可以将其理解为资源对象的"大总管"。
- Kubernetes Scheduler（kube-scheduler）：负责资源调度（Pod调度）的进程，相当于公交公司的"调度室"。
- Etcd：高可用的键-值存储系统组件，Kubernetes使用该组件来存储各个资源的状态，从而实现RESTful API的功能。

Master节点运行的组件主要负责整个Kubernetes集群的控制和管理，包括处理API请求、资源调度、控制器管理和存储集群状态等核心功能。了解并合理配置Master节点及其组件有助于优化Kubernetes集群的性能和稳定性。

1.2.2 Node节点运行的组件

除Master外，Kubernetes集群中的其他机器被称为Node，有时也被称为Worker。在早期版本中，

Node还被称为Minion。与Master一样，Node可以是一台物理主机，也可以是一台虚拟机。Node是Kubernetes集群中的工作负载节点，每个Node都会被Master分配一些工作负载（Docker容器），当某个Node宕机时，其上的工作负载会被Master自动转移到其他节点上。

Node上运行着以下关键组件。

- Kubelet：负责管控容器，Kubelet会从Kubernetes API Server接收Pod的创建请求，启动和停止容器，监控容器运行状态并汇报给Kubernetes API Server。同时与Master密切协作，实现集群管理的基本功能。
- kube-proxy：实现Kubernetes Service的通信与负载均衡机制的重要组件。负责为Pod创建代理服务，Kubernetes Proxy会从Kubernetes API Server获取所有的Service信息，并根据Service的信息创建代理服务，实现Service到Pod的请求路由和转发，从而实现Kubernetes层级的虚拟转发网络。
- Docker Engine（docker）：Docker引擎，负责本机的容器创建和管理工作。

Node节点作为Kubernetes的工作节点，主要负责运行和管理容器化的应用程序，这些节点承担着具体执行工作的任务，是集群的核心执行力量。Node节点上的组件主要用于承载和管理Kubernetes集群中的实际工作负载，包括运行容器、资源分配、节点状态管理等。同样，合理配置Node节点及其组件对于优化Kubernetes集群的性能和稳定性十分重要。

1.3 Kubernetes集群的搭建与配置

在开始学习Kubernetes之前，我们需要准备好集群环境。Kubernetes常用的安装方法包括Kubeadm和二进制安装。Kubeadm和二进制安装可以提供更稳定、更适合生产环境的Kubernetes。本节将以CentOS 7.9操作系统为例，介绍如何通过Kubeadm工具安装Kubernetes。对于Kubernetes运维人员来说，熟练搭建Kubernetes集群是一项必备技能。

1.3.1 Kubeadm 工具介绍

Kubeadm是一个工具，它提供了kubeadm init和kubeadm join这两个命令作为快速创建Kubernetes集群的最佳实践。

Kubeadm通过执行相应的命令来启动和运行必要的操作来启动和运行一个最小可用的集群。它被故意设计为只关心启动集群，而不是之前的节点准备工作。同样，安装各种有价值的插件，例如Kubernetes Dashboard、监控解决方案以及特定云提供商的插件，也不属于它的职责范围。

相反，我们期望由一个基于Kubeadm从更高层设计的更加合适的工具来做这些事情；并且，理想情况下，使用Kubeadm作为所有部署的基础将会使得创建一个符合期望的集群变得容易。

常用命令如下。

- kubeadm init：启动一个Kubernetes主节点。
- kubeadm join：启动一个Kubernetes工作节点并将其加入集群中。
- kubeadm upgrade：更新一个Kubernetes集群到新版本。
- kubeadm token：管理kubeadm join使用的令牌。
- kubeadm reset：还原kubeadm init或者kubeadm join对主机所做的任何更改。
- kubeadm version：打印Kubeadm版本。
- kubeadm config：如果使用v1.8.x以下版本的Kubeadm初始化集群，则需要对集群进行一些配置以便使用kubeadm upgrade命令。

安装Kubeadm需要手动安装Kubelet和Kubectl，因为Kubeadm是不会安装和管理这两个组件的。

1.3.2 基础环境配置

集群搭建开始时，首先需要进行基础环境配置，例如修改主机名、修改各节点的Hosts文件等，具体需要完成的工作如下。

该集群我们规划为4台服务器，即4个节点，分别为k8s-master、k8s-node1、k8s-node2和k8s-node3，每台4核CPU、8GB内存，硬盘为100GB。各节点详细配置如下：

```
k8s-master 192.168.1.10  4核/8G/100GB  Centos7.9-X86_64
k8s-node1  192.168.1.11  4核/8G/100GB  Centos7.9-X86_64
k8s-node2  192.168.1.12  4核/8G/100GB  Centos7.9-X86_64
k8s-node3  192.168.1.13  4核/8G/100GB  Centos7.9-X86_64
```

各节点修改主机名：

```
hostnamectl --static set-hostname k8s-master
hostnamectl --static set-hostname k8s-node1
hostnamectl --static set-hostname k8s-node2
hostnamectl --static set-hostname k8s-node3
```

各节点修改Hosts文件：

```
cat <<EOF > /etc/hosts
127.0.0.1   localhost localhost.localdomain localhost4 localhost4.localdomain4
::1         localhost localhost.localdomain localhost6 localhost6.localdomain6
192.168.1.10  k8s-master
192.168.1.11  k8s-node1
192.168.1.12  k8s-node2
192.168.1.13  k8s-node3
EOF
```

各节点修改DNS：

```
cat <<EOF /etc/resolv.conf
nameserver 114.114.114.114
```

```
nameserver 8.8.8.8
EOF
```

各节点关闭防火墙：

```
systemctl disable firewalld && systemctl stop firewalld
```

该命令用于在Linux系统中禁用和停止firewalld服务。

- **systemctl disable firewalld**：这个命令的作用是将firewalld服务设置为开机不自启，即在系统启动时不会自动启动firewalld服务。这样可以防止firewalld服务在系统启动过程中干扰集群通信，从而避免可能出现的问题。
- **systemctl stop firewalld**：这个命令的作用是立即停止正在运行的firewalld服务。如果firewalld服务当前正在运行，执行此命令后，firewalld服务将立即停止运行。

各节点关闭SELinux：

```
sed -i 's/enforcing/disabled/' /etc/selinux/config && setenforce 0
```

这个命令用于在Linux系统中禁用SELinux（Security-Enhanced Linux）。

- **sed -i 's/enforcing/disabled/' /etc/selinux/config**：这个命令的作用是在/etc/selinux/config文件中查找所有enforcing并替换为disabled。-i选项表示直接修改文件内容，而不是输出到标准输出。这样，SELinux将被设置为禁用模式。
- **setenforce 0**：这个命令的作用是将当前的SELinux策略设置为Permissive模式。在这种模式下，SELinux不会阻止任何操作，但会记录违反策略的操作。这相当于临时禁用SELinux，直到下一次系统重启。

各节点关闭swap分区：

```
swapoff -a && sed -i 's/.*swap.*/#&/' /etc/fstab
```

在Kubernetes集群中，关闭交换分区可以确保内存不被交换出去，避免因为内存不足导致的应用程序崩溃和节点异常。因为Kubernetes会在每个节点上部署多个容器，如果节点上的应用程序占用的内存超过了可用内存，操作系统就会将部分内存换出到硬盘，从而降低容器的性能和稳定性。因此，在部署Kubernetes集群之前，需要关闭交换分区。可以通过命令swapoff -a来关闭交换分区。

各节点加载模块：

```
modprobe br_netfilter && lsmod | grep br_netfilter
```

这里使用modprobe命令加载br_netfilter模块，并使用lsmod命令列出当前加载的所有模块，然后通过grep命令过滤出包含br_netfilter关键字的行。如果输出结果中包含br_netfilter，则说明该模块已经成功加载。

各节点还需要启用网络桥接模式下的iptables包过滤功能；同时，在安装和配置Kubernetes集群时，需要开启一些内核参数以确保网络功能的正常运行。

```
cat > /etc/sysctl.conf <<EOF
net.bridge.bridge-nf-call-ip6tables = 1
net.bridge.bridge-nf-call-iptables = 1
net.bridge.bridge-nf-call-arptables=1
net.ipv4.ip_forward = 1
EOF
```

参数解析：

- net.bridge.bridge-nf-call-ip6tables = 1：允许iptables对IPv6数据包进行处理。
- net.bridge.bridge-nf-call-iptables = 1：允许iptables对桥接数据包进行处理，这是容器之间进行通信所必需的。
- net.bridge.bridge-nf-call-arptables = 1：允许iptables对ARP帧在通过网桥时传递给arptables进行处理。
- net.ipv4.ip_forward = 1：允许在Linux内核中开启IP转发功能，使得数据包可以在不同网络之间进行路由。

各节点加载配置生效：

```
sysctl -p /etc/sysctl.conf
```

此处使用sysctl命令读取/etc/sysctl.conf文件中的配置信息，并将其应用到系统中。这样可以确保系统的配置与配置文件中指定的一致。如果需要修改系统配置，则可以编辑该文件并重新运行该命令来使更改生效。

各节点安装epel-release源：

```
yum install epel-release -y
```

EPEL是一个由Fedora项目提供的第三方软件包存储库，它包含许多在默认的RHEL软件源中不可用的额外软件包。通过安装epel-release软件包，可以将EPEL仓库添加到系统的软件源列表中，从而可以使用yum命令从该仓库安装和更新软件包。-y选项表示自动回答yes，无须用户交互即可完成安装。

各节点安装基础软件：

```
yum install wget yum-utils device-mapper-persistent-data lvm2 ebtables ethtool conntrack-tools -y
```

该命令从默认的软件源中安装以下软件包。

- wget：一个常用的文件下载工具。
- yum-utils：提供一些有用的工具和实用程序，例如yum-config-manager和yumdownloader。

- device-mapper-persistent-data：设备映射器持久性数据（Device Mapper Persistent Data），这是一个用于管理存储设备的Linux内核组件。
- lvm2：逻辑卷管理（Logical Volume Manager），这是一个用于管理存储设备的Linux内核组件。
- ebtables：扩展的桥接表（Extended Berkeley Packet Filter）是一个用于管理以太网桥的网络过滤规则的工具。
- ethtool：一个用于查询和修改网络接口配置的命令行工具。
- conntrack-tools：连接跟踪工具集，包括conntrack、libnetfilter_cttimeout和libnetfilter_cthelper等组件。
- -y：表示自动回答yes，无须用户交互即可完成安装。

各节点添加阿里yum源：

```
wget -O /etc/yum.repos.d/CentOS-Base.repo http://mirrors.aliyun.com/repo/Centos-7.repo && yum clean all && yum makecache
```

该命令将系统的软件源更换为阿里云镜像站点提供的CentOS 7软件源，从而加速软件包的下载速度。具体来说，它执行以下操作：

- 使用wget命令从阿里云镜像站点下载CentOS 7的软件源配置文件，并将其保存到/etc/yum.repos.d/目录下的CentOS-Base.repo文件中。
- 使用yum clean all命令清除所有已缓存的软件包和元数据。
- 使用yum makecache命令重新生成缓存的软件包和元数据，以便系统能够正确识别新的软件源。

各节点添加Docker阿里yum源：

```
wget http://mirrors.aliyun.com/docker-ce/linux/centos/docker-ce.repo -O /etc/yum.repos.d/docker-ce.repo && yum makecache fast
```

该命令将使用阿里云镜像站点提供的Docker CE软件源，从而加速Docker CE软件包的下载速度。具体来说，它执行以下操作：

- 使用wget命令从阿里云镜像站点下载Docker CE的软件源配置文件，并将其保存到/etc/yum.repos.d/目录下的docker-ce.repo文件中。

各节点配置Kubernetes国内阿里yum源：

```
cat <<EOF > /etc/yum.repos.d/kubernetes.repo
[kubernetes]
name=Kubernetes
baseurl=https://mirrors.aliyun.com/kubernetes/yum/repos/kubernetes-el7-x86_64/
enabled=1
gpgkey=https://mirrors.aliyun.com/kubernetes/yum/doc/yum-key.gpg
```

```
https://mirrors.aliyun.com/kubernetes/yum/doc/rpm-package-key.gpg
EOF
```

各节点生成缓存及更新系统：

```
yum makecache fast -y && yum update -y
```

该命令用于在Linux系统中更新软件包，使用的是yum包管理器。

- yum makecache fast -y：这个命令的作用是创建一个本地的yum缓存，其中包含所有可用的软件包和版本信息。makecache fast选项表示使用快速模式来创建缓存，这通常比默认模式更快，但可能会稍微降低缓存的准确性。-y选项表示在执行过程中自动回答"是"，即不需要用户手动确认。
- yum update -y：这个命令的作用是更新系统中所有已安装的软件包到最新版本。update选项表示更新所有软件包，-y选项表示在执行过程中自动回答"是"，即不需要用户手动确认。

各节点安装指定版本docker-ce：

```
yum install docker-ce-19.03.9-3.el7 -y
```

该命令的作用是安装版本为19.03.9-3.el7的Docker CE。install选项表示安装软件包，docker-ce-19.03.9-3.el7是软件包的名称和版本。-y选项表示在执行过程中自动回答"是"，即不需要用户手动确认。

各节点启动Docker并设置为开机启动：

```
systemctl enable docker --now
```

该命令用于在Linux系统中设置Docker服务开机自启动，并立即启动Docker服务。

在Linux系统中，Docker是一个流行的容器化平台，它允许用户在隔离的容器中运行和管理应用程序。使用systemctl enable docker命令可以将Docker服务设置为开机自启动，确保每次系统重启后Docker会自动运行，无须手动启动。这个命令适用于使用Systemd作为初始化系统的Linux发行版，如Ubuntu 16.04及以后版本、Debian 8及以后版本和CentOS 7及以后版本等。

具体来说，systemctl enable docker命令会创建相应的Systemd链接，指向Docker的单元文件（通常位于/usr/lib/systemd/system或/etc/systemd/system目录下）。这样，在下一次启动时，Systemd会读取这些链接并自动启动Docker服务。

此外，--now参数则告诉Systemd立即启动Docker服务，而不是等待下一次系统启动。这意味着执行该命令后，Docker服务将立即被激活并运行。通过这种方式，用户可以确保Docker服务在当前会话中以及未来每次系统启动时都有效运行。

各节点配置镜像加速器、日志驱动以及Cgroup驱动：

```
sudo tee /etc/docker/daemon.json <<-'EOF'
{
    "registry-mirrors": ["https://nty7c4os.mirror.aliyuncs.com"],
    "live-restore":true,
    "exec-opts":["native.cgroupdriver=systemd"],
    "log-driver":"json-file",
    "log-opts":{
        "max-size":"100m",
        "max-file":"3"
    }
}
EOF
```

上述命令用于在Linux系统中配置Docker守护进程的设置。具体来说，它使用sudo tee命令将一个JSON格式的配置写入/etc/docker/daemon.json文件中。每个选项的含义如下。

- registry-mirrors：这是一个数组，指定了Docker镜像仓库的镜像地址。在这个例子中，使用了阿里云的镜像加速器地址https://nty7c4os.mirror.aliyuncs.com，以提高拉取镜像的速度。
- live-restore：这个选项设置为true，表示Docker守护进程在容器退出时尝试恢复容器的状态，以便下一次启动时能够继续运行。
- exec-opts：这是一个数组，用于指定额外的参数传递给容器运行时。在这个例子中，设置了native.cgroupdriver=systemd，这意味着Docker将使用Systemd作为其Cgroup驱动。
- log-driver：指定Docker日志驱动程序的名称。在这个例子中，使用了json-file驱动程序，它会将日志输出为JSON格式的文件。
- log-opts：这是一个对象，用于指定日志驱动程序的选项。在这个例子中，设置了max-size为100MB和max-file为3，意味着日志文件的最大大小为100MB，并且最多保留3个日志文件。

通过执行这个命令，可以修改Docker守护进程的配置，以适应特定的需求或优化性能。

热加载配置和重启Docker：

```
systemctl daemon-reload && systemctl restart docker
```

上述命令用于重新加载Docker服务的配置文件并重启Docker服务。

- systemctl daemon-reload命令会重新读取并应用配置文件中定义的设置，使更改立即生效而无须重启服务。这对于在不中断服务的情况下进行配置更改非常有用。例如，假设我们有一个名为docker.service的服务，并且修改了其配置文件。如果我们想要使这些更改生效，而不想停止和重新启动整个服务，可以使用daemon-reload命令。
- systemctl restart docker命令用于立即重启Docker服务。这意味着先停止当前的Docker服务进程，然后根据新的配置文件启动一个新的进程。通过这种方式，用户可以确保Docker服务在当前会话中以及未来每次系统启动时都有效运行。

1.3.3 升级内核版本

在基础环境配置完成后,还需要升级Linux内核为长期稳定版本。升级的原因是:在一个Node上,经过测试发现,使用3.10版本的内核时,当并行创建7个任务并同时触发内存溢出时,会导致内核锁死,因此需要将内核版本升级到最新版本。

具体步骤如下:

01 各节点导入ELRepo公钥:

```
rpm -import https://www.elrepo.org/RPM-GPG-KEY-elrepo.org
```

这个命令用于导入一个GPG密钥,以便在安装或更新来自ELrepo存储库的软件包时进行签名验证。具体来说,它使用rpm命令的-import选项来导入指定的GPG密钥。在这个例子中,GPG密钥的URL是https://www.elrepo.org/RPM-GPG-KEY-elrepo.org。

02 各节点安装ELRepo仓库:

```
rpm -Uvh https://www.elrepo.org/elrepo-release-7.0-3.el7.elrepo.noarch.rpm
```

这个命令用于在Linux系统中安装或更新ELRepo存储库的软件包。具体来说,它使用rpm命令的-Uvh选项来升级或安装指定的软件包。在这个例子中,要安装的软件包的URL是https://www.elrepo.org/elrepo-release-7.0-3.el7.elrepo.noarch.rpm。其中,-U表示升级或安装软件包,-v表示显示详细信息,-h表示显示进度条。

03 各节点安装kernel-ml内核,lt表示长期维护版本,ml表示长期稳定版本:

```
yum -y --enablerepo=elrepo-kernel install kernel-ml.x86_64 kernel-ml-devel.x86_64
```

使用yum命令的-y选项来自动回答yes,即在安装过程中不会出现任何提示,直接进行安装。使用--enablerepo=elrepo-kernel选项来指定要使用的存储库名称为elrepo-kernel。要安装的软件包的名称是kernel-ml.x86_64和kernel-ml-devel.x86_64,分别表示内核映像和内核开发头文件。

04 各节点查看启动器:

```
awk -F\' '$1=="menuentry " {print i++ " : " $2}' /etc/grub2.cfg
```

这个命令用于列出Grub引导加载程序的菜单项。具体来说,它使用awk命令来处理/etc/grub2.cfg文件中的内容。-F\'表示将单引号作为字段分隔符,$1=="menuentry "表示选择包含"menuentry "字符串的行,{print i++ " : "$2}表示打印出该行的第二个字段(即菜单项名称)以及一个自增计数器i的值。最终输出的结果是一个数字和菜单项名称的列表。

05 各节点设置默认启动为新内核:

```
grub2-set-default 0
```

这个命令用于设置Grub引导加载程序的默认启动项。具体来说，它使用grub2-set-default命令来设置默认启动项的编号。在这个例子中，0表示将默认的启动项设置为启动菜单的第一个菜单项。如果需要设置其他菜单项为默认启动项，则可以将0替换为相应的数字。

06 各节点重启系统：

```
reboot
```

07 各节点安装ipvs和管理工具：

```
yum install ipset ipvsadm -y
```

参数解析：

- ipset ipvsadm：这是要安装的软件包的名称，ipset是一个用于管理IP集合的工具，而ipvsadm是一个用于配置Linux内核中的IP虚拟服务器（IP Virtual Server，IPVS）的工具。
- -y：这是一个选项，表示在安装过程中自动回答yes，即不需要用户确认是否继续安装。

08 各节点创建模块加载脚本：

```
vim /etc/sysconfig/modules/ipvs.modules
# 脚本内容如下
#!/bin/bash
modprobe -- ip_vs
modprobe -- ip_vs_rr
modprobe -- ip_vs_wrr
modprobe -- ip_vs_sh
modprobe -- nf_conntrack          # 内核版本如果为3.10，则使用nf_conntrack_ipv4模块
modprobe -- br_netfilter
```

该命令将创建一个模块脚本，这些脚本在系统启动时由Systemd或其他初始化系统调用，以确保特定的内核模块在系统启动过程中被加载。这样做可以自动化配置过程，确保关键的内核模块在系统启动时就已经准备好，无须手动干预。

参数解析：

- br_netfilter叫作透明防火墙（Transparent Firewall），又称桥接模式防火墙（Bridge Firewall）。简单来说，就是在网桥设备上加入防火墙功能。透明防火墙具有部署能力强、隐蔽性好、安全性高的优点。开启ip6tables和iptables需要加载透明防火墙。可以使用modprobe命令加载Linux机器的内核模块。
- modprobe命令用于加载内核模块br_netfilter。该模块可用于Linux系统中的网桥（Bridge）和网络过滤（Netfilter）的交互。它实现了一个netfilter hook函数，使其能够处理网桥的数据包。当该模块被加载时，它将被插入内核中，使得网络数据包可以正确地被过滤和转发。

09 各节点赋值权限并加载ipvs模块：

```
chmod 755 /etc/sysconfig/modules/ipvs.modules && bash /etc/sysconfig/modules/
ipvs.modules
```

该命令用于更改ipvs.modules文件的权限并执行该文件。具体来说，它使用chmod命令来更改文件的权限为755，即用户具有读、写和执行权限，组和其他用户具有读和执行权限。然后，它使用bash命令来执行/etc/sysconfig/modules/ipvs.modules文件中的命令。这些命令通常用于加载或卸载ipvsadm模块，以便在系统中启用或禁用IP虚拟服务器（IPVS）功能。

至此，我们完成了Linux系统内核的升级。

1.3.4　三大组件的安装

各节点执行如下命令，以查看Kubernetes源的当前状态：

```
yum repolist all | grep kubernetes
```

该命令用于列出所有已启用的Yum存储库，并使用grep命令过滤出包含kubernetes关键字的行。

参数解析：

- yum repolist all命令列出所有已启用的Yum存储库，然后通过管道将输出传递给grep命令进行过滤。
- grep命令使用-k选项只输出包含kubernetes关键字的行。最终输出的结果是一个包含kubernetes相关存储库名称和版本号的列表。

如果源状态为disable，则执行如下命令使Kubernetes源生效：

```
yum-config-manager --enable Kubernetes
```

使用yum-config-manager命令的--enable选项来启用指定的存储库。在这个例子中，要启用的存储库的名称是Kubernetes。如果需要启用其他名称的存储库，则可以将Kubernetes替换为相应的名称。

各节点安装指定版本Kubelet、Kubeadm和Kubectl：

```
yum -y install kubelet-1.23.17 kubeadm-1.23.17 kubectl-1.23.17
--disableexcludes=kubernetes
```

该命令用于在Linux系统中安装或更新Kubernetes相关的软件包。

使用--disableexcludes=kubernetes选项可以禁用排除特定仓库中的软件包。安装的软件包的名称是kubelet-1.23.17、kubeadm-1.23.17和kubectl-1.23.17。这些软件包分别表示Kubernetes节点、集群初始化工具和命令行工具。如果需要安装其他版本的软件包，则可以将版本号替换为相应的值。

各节点查看配置文件目录：

```
# rpm -ql kubelet
/etc/kubernetes/manifests                    # 清单目录
/etc/sysconfig/kubelet                       # 配置文件
```

```
/usr/bin/kubelet                          # 主程序
/usr/lib/systemd/system/kubelet.service   # 服务启动文件
```

各节点设置Kubelet开机自启：

```
systemctl enable kubelet
```

至此，完成了三大组件Kubelet、Kubeadm、Kubectl的安装。

1.3.5 集群镜像处理

因为Kubernetes（K8s）中的组件所需要的镜像因网络限制无法在国内下载，通常采用一些策略来从国外的镜像仓库中获取这些镜像。我们需要使用VPN或代理来绕过网络限制。本小节将采用从国内阿里云镜像仓库进行拉取以解决镜像问题。具体步骤如下：

01 在k8s-master节点查看Kubernetes需要哪些镜像：

```
[root@k8s-master ~]# kubeadm config images list
registry.k8s.io/kube-apiserver:v1.23.17
registry.k8s.io/kube-controller-manager:v1.23.17
registry.k8s.io/kube-scheduler:v1.23.17
registry.k8s.io/kube-proxy:v1.23.17
registry.k8s.io/pause:3.6
registry.k8s.io/etcd:3.5.6-0
registry.k8s.io/coredns/coredns:v1.8.6
```

该命令使用Kubeadm工具列出了Kubernetes集群所需的镜像列表。这些镜像包括Kubernetes的各个组件，如API Server、Controller Manager、Scheduler、Kube-Proxy、Pause容器等。同时，还包括Etcd和CoreDNS等其他组件的镜像。这些镜像将被用于部署Kubernetes集群的各个节点。

02 各节点下载所需镜像：

```
# 查看版本号并替换为阿里云镜像仓库源下载
[root@k8s-master ~]# kubeadm config images list | sed -e 's/^/docker pull /g' -e 's#registry.k8s.io#registry.aliyuncs.com/google_containers#g' | sh -x
docker pull registry.aliyuncs.com/google_containers/coredns:v1.8.6
```

该命令使用阿里云镜像仓库源下载Kubernetes集群所需的镜像。首先，通过kubeadm config images list命令列出了Kubernetes集群所需的镜像列表，然后使用sed命令将镜像名称中的registry.k8s.io替换为阿里云的镜像仓库地址registry.aliyuncs.com/google_containers，并在每个镜像名称前添加了docker pull命令。最后，使用sh -x命令执行生成的命令行脚本，以下载所需的镜像。

> **注意** 在Kubernetes（K8s）中，Node节点加入集群时，涉及多个镜像的拉取和使用，其中包括registry.k8s.io/pause:3.6和registry.k8s.io/kube-proxy:v1.23.17。

1.3.6 集群初始化

在k8s-master节点使用Kubeadm初始化集群：

```
[root@k8s-master ~]# kubeadm init  --kubernetes-version=v1.23.17 --image-repository
registry.aliyuncs.com/google_containers --apiserver-advertise-address=192.168.1.10
--pod-network-cidr=10.244.0.0/16 --service-cidr=10.96.0.0/12
--ignore-preflight-errors=Swap --dry-run
```

参数解析：

- --kubernetes-version：指定Kubernetes集群版本。
- --image-repository registry.aliyuncs.com/google_containers：指定镜像源地址。
- --apiserver-advertise-address：指定集群ApiServer地址，通常为master节点IP。
- --pod-network-cidr：指定Pod分配的IP地址及范围。
- --service-cidr：指定Service分配的IP地址及范围。
- --ignore-preflight-errors：使用--ignore-preflight-errors选项时，用户需要明确指定要忽略的错误类型或错误代码。
- --dry-run：试运行不应用任何变化，测试没问题后将该参数去掉，执行完成初始化集群即可。

1.3.7 Node 节点加入集群

上述集群初始化成功后，显示如图1-1所示。

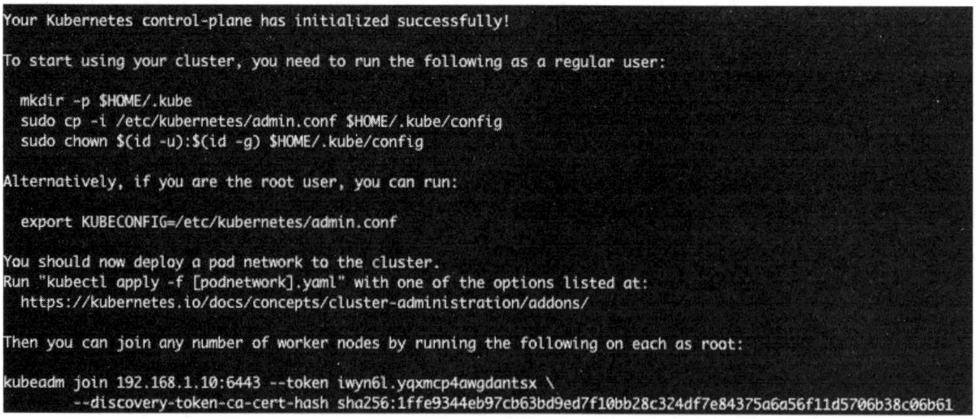

```
Your Kubernetes control-plane has initialized successfully!

To start using your cluster, you need to run the following as a regular user:

  mkdir -p $HOME/.kube
  sudo cp -i /etc/kubernetes/admin.conf $HOME/.kube/config
  sudo chown $(id -u):$(id -g) $HOME/.kube/config

Alternatively, if you are the root user, you can run:

  export KUBECONFIG=/etc/kubernetes/admin.conf

You should now deploy a pod network to the cluster.
Run "kubectl apply -f [podnetwork].yaml" with one of the options listed at:
  https://kubernetes.io/docs/concepts/cluster-administration/addons/

Then you can join any number of worker nodes by running the following on each as root:

kubeadm join 192.168.1.10:6443 --token iwyn6l.yqxmcp4awgdantsx \
    --discovery-token-ca-cert-hash sha256:1ffe9344eb97cb63bd9ed7f10bb28c324df7e84375a6a56f11d5706b38c06b61
```

图 1-1　集群初始化成功后的效果

在k8s-master节点初始化配置文件：

```
# 在当前用户家目录下创建一个目录.kube，用来存放安全上下文config文件
[root@k8s-master ~]# mkdir -p $HOME/.kube
# 将配置文件admin.conf复制到当前用户的家目录下的.kube目录中，并将它重命名为config
```

```
[root@k8s-master ~]# sudo cp -i /etc/kubernetes/admin.conf $HOME/.kube/config
# 将当前用户的家目录下的.kube目录中的config文件所有者更改为当前用户
[root@k8s-master ~]# sudo chown $(id -u):$(id -g) $HOME/.kube/config
```

在各Node节点上执行加入集群命令：

```
kubeadm join 192.168.1.10:6443 --token iwyn6l.yqxmcp4awgdantsx \
        --discovery-token-ca-cert-hash
sha256:1ffe9344eb97cb63bd9ed7f10bb28c324df7e84375a6a56f11d5706b38c06b61
```

参数解析：

- kubeadm join：一个Kubeadm的子命令，用于将节点加入集群。
- 192.168.1.10:6443：Kubernetes主节点的地址，即新节点需要连接的主节点。这个地址包含端口号6443，它是Kubernetes API Server的默认端口号。
- --token iwyn6l.yqxmcp4awgdantsx：一个用于授权新节点加入集群的令牌，它是由Kubernetes主节点生成的。这个特定的令牌是iwyn6l.yqxmcp4awgdantsx。
- --discovery-token-ca-cert-hash sha256：一个证书散列值，它用于验证节点是否能够加入Kubernetes集群。在这个例子中，证书散列值是1ffe9344eb97cb63bd9ed7f10bb28c324df7e84375a6a56f11d5706b38c06b61。

节点加入成功后如图1-2所示。

```
[root@k8s-node1 ~]# kubeadm join 192.168.1.10:6443 --token iwyn6l.yqxmcp4awgdantsx \
>       --discovery-token-ca-cert-hash sha256:1ffe9344eb97cb63bd9ed7f10bb28c324df7e84375a6a56f11d5706b38c06b61
[preflight] Running pre-flight checks
[preflight] Reading configuration from the cluster...
[preflight] FYI: You can look at this config file with 'kubectl -n kube-system get cm kubeadm-config -o yaml'
[kubelet-start] Writing kubelet configuration to file "/var/lib/kubelet/config.yaml"
[kubelet-start] Writing kubelet environment file with flags to file "/var/lib/kubelet/kubeadm-flags.env"
[kubelet-start] Starting the kubelet
[kubelet-start] Waiting for the kubelet to perform the TLS Bootstrap...

This node has joined the cluster:
* Certificate signing request was sent to apiserver and a response was received.
* The Kubelet was informed of the new secure connection details.

Run 'kubectl get nodes' on the control-plane to see this node join the cluster.
```

图 1-2　集群加入节点后的效果

查看节点状态，如图1-3所示，K8s集群因网络插件未安装，所以显示节点状态为NotReady。

```
[root@k8s-master ~]# kubectl get nodes -o wide
NAME         STATUS     ROLES                  AGE    VERSION    INTERNAL-IP    EXTERNAL-IP   OS-IMAGE                KERNEL-VERSION                CONTAINER-RUNTIME
k8s-master   NotReady   control-plane,master   174m   v1.23.17   192.168.1.10   <none>        CentOS Linux 7 (Core)   6.8.9-1.el7.elrepo.x86_64     docker://19.3.9
k8s-node1    NotReady   <none>                 114s   v1.23.17   192.168.1.11   <none>        CentOS Linux 7 (Core)   6.8.9-1.el7.elrepo.x86_64     docker://19.3.9
k8s-node2    NotReady   <none>                 96s    v1.23.17   192.168.1.12   <none>        CentOS Linux 7 (Core)   6.8.9-1.el7.elrepo.x86_64     docker://19.3.9
k8s-node3    NotReady   <none>                 90s    v1.23.17   192.168.1.13   <none>        CentOS Linux 7 (Core)   6.8.9-1.el7.elrepo.x86_64     docker://19.3.9
```

图 1-3　当前各节点的状态

上述结果说明如下。

- NAME：Kubernetes控制节点名称。

- STATUS：集群状态，此时集群状态还是NotReady，因为没有安装网络插件。
- ROLES：集群角色，其中k8s-master指该节点作为Kubernetes控制平面。
- AGE：集群创建的时间。
- VERSION：Kubernetes集群版本。
- EXTERNAL-IP：Kubernetes节点的公共IP地址。
- OS-IMAGE：Kubernetes集群节点的系统镜像版本。
- KERNEL-VERSION：Kubernetes集群节点的系统内核版本。
- CONTAINER-RUNTIME：容器运行时，用于管理容器，以便能够将容器化的应用程序部署和运行在集群中。Docker和Containerd都是常用的容器运行时。

1.3.8　Calico 网络插件安装

通过前几节，我们已经把Kubernetes安装成功了，但是Kubernetes的状态一直是NotReady，说明没有安装网络插件，本小节将带领读者安装网络插件Calico。

Calico是一种容器之间互通的网络方案，基于BGP的纯三层网络架构。它能够与OpenStack、Kubernetes、AWS、GCE等云平台良好集成。在虚拟化平台中，比如OpenStack、Docker等，需要实现跨主机互连，并对容器进行隔离控制。然而，大多数虚拟化平台使用二层隔离技术来实现容器网络，这些技术存在一些弊端，比如需要依赖VLAN、Bridge和Tunnel（隧道）等技术。其中，Bridge带来了复杂性，而VLAN隔离和隧道在拆包或加包头时会消耗更多的资源，并且对物理环境也有要求。随着网络规模的增大，整体会变得越加复杂。

在Kubernetes集群中，如果没有安装网络插件，那么Kubernetes集群就不会正常工作，无法跨节点通信，也无法对Pod进行IP分配。Kubernetes支持多种网络插件，Calico是目前使用最多的一个，因为Calico效率高，还支持网络隔离，应用范围更广。

下面介绍Calico网络插件的安装步骤。

01 下载Calico网络插件资源清单：

```
[root@k8s-master ~]# wget https://docs.projectcalico.org/manifests/calico.yaml
--no-check-certificate
```

该命令从指定的 https://docs.projectcalico.org/manifests/calico.yaml URL下载一个名为calico.yaml的文件，并将其保存到当前目录中。同时使用--no-check-certificate选项来禁用SSL证书验证。如果下载成功，该文件将被保存到当前目录下的calico.yaml文件中。

02 修改Calico资源清单文件：

```
#搜索IPV4POOL，找到如下内容，取消注释，改成想要设置的Pod IP段，通常设置为10.244.0.0/16
- name: CALICO_IPV4POOL_CIDR
  value: "10.244.0.0/16"
```

03 应用Calico资源清单：

```
[root@k8s-master ~]# kubectl apply -f calico.yaml
clusterrolebinding.rbac.authorization.k8s.io/calico-kube-controllers created
clusterrolebinding.rbac.authorization.k8s.io/calico-node created
...
daemonset.apps/calico-node created
deployment.apps/calico-kube-controllers created
```

参数解析：

- kubectl apply：用于通过YAML配置文件来创建或更新Kubernetes资源。
- -f calico.yaml：将指定的YAML文件应用于当前的Kubernetes环境。
- yaml：YAML（Yet Another Markup Language）资源清单是Kubernetes中用于定义和描述资源对象配置的主要格式。与JSON相比，YAML更加简洁、易读，并且支持注释，这使得它在Kubernetes的配置管理中备受欢迎。

04 查看Kubernetes节点状态，网络插件安装后，节点状态变为Ready：

```
[root@k8s-master ~]# kubectl get nodes
NAME         STATUS   ROLES                  AGE    VERSION
k8s-master   Ready    control-plane,master   4h38m  v1.23.17
k8s-node1    Ready    <none>                 105m   v1.23.17
k8s-node2    Ready    <none>                 105m   v1.23.17
k8s-node3    Ready    <none>                 105m   v1.23.17
```

参数解析：

- NAME：节点的名称，即在Kubernetes集群中给该节点分配的唯一标识符。
- STATUS：节点的状态，表示该节点在Kubernetes集群中是否已准备好接受工作负载。在本例中，所有节点的状态都为Ready，表示它们都已经准备好。
- ROLES：节点的角色，即该节点在Kubernetes集群中所扮演的角色。在本例中，k8s-master节点的角色为control-plane和master，表示它是Kubernetes控制平面的主节点；而k8s-node1、k8s-node2和k8s-node3节点的角色为<none>，表示它们不是Kubernetes控制平面的节点，代表工作节点。
- AGE：节点的时间，即该节点在Kubernetes集群中运行的时间。
- VERSION：节点使用的Kubernetes版本，即它所运行的Kubernetes软件的版本号。

1.3.9　Metrics-Server 服务的安装

我们已经安装完成Kubernetes集群，并且各节点的状态为Ready。接下来将带领读者了解并安装Metrics-Server服务，旨为让我们的集群具有更好的资源监控功能。

Metrics-Server是Kubernetes集群中的核心组件之一，主要负责从Kubelet收集资源使用情况的指标,并对这些监控数据进行聚合(依赖于kube-aggregator)。它通过Kubernetes Apiserver的Metrics API (/apis/metrics.k8s.io)对外公开这些指标,但仅保留最新的指标数据(如CPU和内存使用情况)。

Metrics-Server的主要作用是为kube-scheduler、HorizontalPodAutoscaler等Kubernetes核心组件,以及kubectl top命令和Dashboard等UI组件提供数据支持。此外,Metrics-Server还支持自定义监控指标,比如通过集成k8s-prometheus-adapter等扩展来实现。

下面介绍Metrics-Server的安装步骤。

01 下载metrics-server资源清单:

```
[root@k8s-master ~]# wget -O metrics-server.yaml
https://github.com/kubernetes-sigs/metrics-server/releases/download/v0.7.0/components.yaml --no-check-certificate
```

该命令从指定的 https://github.com/kubernetes-sigs/metrics-server/releases/download/v0.7.0/components.yaml URL下载一个名为metrics-server.yaml的文件,并将其保存到当前目录中。同时使用--no-check-certificate选项来禁用SSL证书验证。如果下载成功,该文件将被保存到当前目录下的metrics-server.yaml文件中。

02 各节点下载镜像:

```
docker pull registry.aliyuncs.com/google_containers/metrics-server:v0.7.0
```

该命令从阿里云的镜像仓库下载一个名为metrics-server:v0.7.0的镜像。如果下载成功,该镜像将被保存到本地Docker镜像库中。

03 修改metrics-server资源清单:

```
    spec:
      containers:
      - args:
        - --cert-dir=/tmp
        - --secure-port=10250
        - --kubelet-preferred-address-types=InternalIP,ExternalIP,Hostname
        - --kubelet-use-node-status-port
        - --metric-resolution=15s
        # 禁用证书验证
        - --kubelet-insecure-tls
        # 修改镜像为阿里云镜像地址
        image: registry.aliyuncs.com/google_containers/metrics-server:v0.7.0
        imagePullPolicy: IfNotPresent
```

04 应用metrics-server资源清单:

```
[root@k8s-master ~]# kubectl apply -f metrics-server.yaml
serviceaccount/metrics-server created
clusterrole.rbac.authorization.k8s.io/system:aggregated-metrics-reader created
```

```
clusterrole.rbac.authorization.k8s.io/system:metrics-server created
rolebinding.rbac.authorization.k8s.io/metrics-server-auth-reader created
clusterrolebinding.rbac.authorization.k8s.io/metrics-server:system:auth-delegator
created
clusterrolebinding.rbac.authorization.k8s.io/system:metrics-server created
service/metrics-server created
deployment.apps/metrics-server created
apiservice.apiregistration.k8s.io/v1beta1.metrics.k8s.io created
```

该命令使用Kubectl工具将metrics-server.yaml文件中定义的资源应用到Kubernetes集群中。具体来说，它使用kubectl apply命令来创建或更新资源，并使用-f选项指定要应用的文件名。根据输出信息，可以看出已经成功创建了以下资源：

- serviceaccount/metrics-server
- clusterrole.rbac.authorization.k8s.io/system:aggregated-metrics-reader
- clusterrole.rbac.authorization.k8s.io/system:metrics-server
- rolebinding.rbac.authorization.k8s.io/metrics-server-auth-reader
- clusterrolebinding.rbac.authorization.k8s.io/metrics-server:system:auth-delegator
- clusterrolebinding.rbac.authorization.k8s.io/system:metrics-server
- service/metrics-server
- deployment.apps/metrics-server
- apiservice.apiregistration.k8s.io/v1beta1.metrics.k8s.io

05 查看节点资源，通过命令kubectl top [资源类型node/pods]验证metrics-server是否安装成功：

```
[root@k8s-master ~]# kubectl top node
NAME         CPU(cores)   CPU%   MEMORY(bytes)   MEMORY%
k8s-master   84m          4%     1424Mi          37%
k8s-node1    73m          3%     1157Mi          30%
k8s-node2    41m          2%     906Mi           23%
k8s-node3    43m          2%     906Mi           23%
```

06 查看Pods资源信息：

```
[root@k8s-master ~]# kubectl top pods -A
NAMESPACE     NAME                                       CPU(cores)   MEMORY(bytes)
kube-system   calico-kube-controllers-64cc74d646-f25m2   1m           20Mi
kube-system   calico-node-cj8mf                          13m          122Mi
kube-system   calico-node-rzg6c                          16m          119Mi
kube-system   calico-node-x5kmj                          12m          119Mi
kube-system   calico-node-xfz8n                          14m          118Mi
kube-system   coredns-6d8c4cb4d-frw8v                    1m           17Mi
kube-system   coredns-6d8c4cb4d-lnfsp                    1m           18Mi
kube-system   etcd-k8s-master                            8m           69Mi
kube-system   kube-apiserver-k8s-master                  24m          315Mi
kube-system   kube-controller-manager-k8s-master         7m           52Mi
```

```
kube-system   kube-proxy-8w9p5                          1m    15Mi
kube-system   kube-proxy-gc8bs                          1m    17Mi
kube-system   kube-proxy-txvzw                          1m    17Mi
kube-system   kube-proxy-vpngb                          1m    15Mi
kube-system   kube-scheduler-k8s-master                 2m    20Mi
kube-system   metrics-server-5cc466bcdc-6kmmc           2m    20Mi
```

该命令用于获取Kubernetes节点的资源使用信息:

- NAME: Kubernetes集群节点名。
- CPU(cores): 节点上当前使用的CPU核心数（以m为单位，1000m等于1核心）。
- CPU%: 节点上CPU的使用率百分比。
- MEMORY(bytes): 节点上当前使用的内存量（以字节为单位）。
- MEMORY%: 节点上内存的使用率百分比。

系统资源数据取值计算方法:

- 输出的CPU和内存数据是近似的，可能不是实时的，因为收集这些指标需要一些时间。
- 输出中的百分比是基于节点的总容量来计算的。例如，如果节点有4个CPU核心和32GiB内存，并且显示CPU使用率为27%，那么这意味着大约有1.08个核心正在被使用。

1.3.10　Kuboard 管理平台安装

在市场上有许多管理Kubernetes集群的工具及Web系统，常用的Web管理系统有官方的Dashboard和第三方的Kuboard。本书将使用Kuboard作为Kubernetes集群的管理平台。

Kuboard是一款基于Kubernetes的开源Web界面管理工具，它可以帮助用户轻松地管理和监控Kubernetes集群，提高开发和运维的效率。Kuboard兼容Kubernetes版本1.13及以上，且每周发布一个beta版本，最长每月发布一个正式版本，经过两年来的不断迭代和优化，已经具备多集群管理、权限管理、监控套件、日志套件等丰富的功能，并且有1000多家企业将Kuboard应用于其生产环境。

下面介绍Kuboard的安装步骤。

01 在k8s-master节点下载资源清单:

```
[root@k8s-master ~]# wget https://addons.kuboard.cn/kuboard/kuboard-v3.yaml
--no-check-certificate
```

该命令从指定的URL下载一个名为kuboard-v3.yaml的文件，如果下载成功，该文件将被保存到当前目录下的kuboard-v3.yaml文件中。

02 应用资源清单:

```
[root@k8s-master ~]# kubectl apply -f kuboard-v3.yaml
namespace/kuboard created
configmap/kuboard-v3-config created
```

```
serviceaccount/kuboard-boostrap created
clusterrolebinding.rbac.authorization.k8s.io/kuboard-boostrap-crb created
daemonset.apps/kuboard-etcd created
deployment.apps/kuboard-v3 created
service/kuboard-v3 created
```

该命令使用Kubectl工具将kuboard-v3.yaml文件中定义的资源应用到Kubernetes集群中。使用-f选项指定要应用的文件名。根据输出信息，可以看出已经成功创建了以下资源：

- namespace/kuboard
- configmap/kuboard-v3-config
- serviceaccount/kuboard-boostrap
- clusterrolebinding.rbac.authorization.k8s.io/kuboard-boostrap-crb
- daemonset.apps/kuboard-etcd
- deployment.apps/kuboard-v3
- service/kuboard-v3

03 查看Kuboard访问端口：

```
[root@k8s-master ~]# kubectl get svc -n kuboard
NAME         TYPE       CLUSTER-IP       EXTERNAL-IP   PORT(S)                                         AGE
kuboard-v3   NodePort   10.105.148.207   <none>        80:30080/TCP,10081:30081/TCP,10081:30081/UDP   13m
```

参数解析：

- NAME：服务的名称，在上面的例子中，服务的名称是kuboard-v3。
- TYPE：服务的类型，常见的类型有ClusterIP、NodePort、LoadBalancer和ExternalName。
- CLUSTER-IP：集群内部IP地址，对于ClusterIP类型的服务，这个IP地址在集群内部是可达的。在上面的例子中，IP地址是10.105.148.207。
- EXTERNAL-IP：集群外部IP地址，对于LoadBalancer或NodePort类型的服务，如果外部IP地址已经分配，它将会显示在这里。但在ClusterIP类型的服务中，这个字段通常是<none>。
- PORT(S)：服务暴露的端口，可以看到端口和协议（TCP或UDP）。在上面的例子中，服务暴露了80端口映射到各节点上的30080端口，并使用TCP协议。
- AGE：服务创建后已经存在的时间。

04 访问Kuboard：http://192.168.1.10:30080。

```
账户：admin
密码：Kuboard123
```

至此，我们已经成功安装了Kuboard，之后就可以使用Kuboard来管理Kubernets集群了。

1.4 本章小结

本章介绍了Kubernetes的基本概念和组件。首先，解释了为什么需要使用Kubernetes来管理和调度容器化应用程序。然后，详细介绍了Kubernetes节点的组件，包括master节点组件和node节点组件。接下来，讲解了如何搭建和配置Kubernetes集群。通过介绍Kubeadm工具、基础环境配置、升级内核版本等，读者可以了解如何搭建一个基本的Kubernetes集群。最后，介绍了如何安装calico网络插件和metrics-server服务，并安装了Kuboard来进行可视化管理。通过本章的学习和实践，读者可以对Kubernetes有一个初步的了解，为后续章节的学习打下基础。

第 2 章

Pod控制器

在Kubernetes中，Pod是应用程序的最小可部署单元，它封装了一个或多个紧密相关的容器，共享存储、网络和资源。然而，仅仅依靠Pod来管理和编排集群中的资源是远远不够的。为了更高效地管理和扩展Pod，Kubernetes引入了Pod控制器。

本章将带领读者深入了解Kubernetes Pod控制器的概念、类型及其工作原理。Pod控制器是Kubernetes集群中负责管理和控制Pod生命周期的组件，它们确保Pod按照预期的状态运行，并在需要时创建、更新或删除Pod。

我们将首先介绍几种常见的Pod控制器，如Replication、ReplicaSet、Deployment、StatefulSet、DaemonSet和Job等，并解释它们各自的使用场景和优势。通过具体的示例和案例，将展示如何使用这些控制器来部署、扩展和管理Pod。此外，我们还将探讨Pod控制器的生命周期管理，包括如何定义Pod的期望状态、如何监控Pod的实际状态，并在状态发生变化时采取相应的措施。通过深入理解Pod控制器的工作机制，读者将能够更好地利用Kubernetes的编排功能，实现高效、可靠的应用程序部署和管理。

2.1 Pod、Kubectl与YAML

在学习控制器之前，本节首先介绍几个重要的知识点，即Pod、Kubectl和YAML，这对后续理解控制器非常重要。

2.1.1 Pod及其操作

1. 什么是Pod

Kubernetes的Pod是由一个或多个容器组成的最小可部署单元,这些容器共享相同的网络命名空间和存储卷,形成逻辑上紧密耦合的单元。

在Kubernetes中,Pod是资源调度和管理的基本单位。Pod可以包含一个或多个容器,这些容器通过localhost互相通信,共享存储卷和网络命名空间。这种设计使得Pod内的容器能够高效地共享数据和通信,从而协同完成特定任务。

每个Pod都有一个唯一的IP地址,这个IP地址在整个Pod生命周期内是不变的。Pod内的容器不会单独获得IP地址,而是通过Pod的IP地址进行通信。这种设计简化了网络配置,使得同一Pod内的容器可以通过localhost以及不同的端口直接进行相互通信。

Pod具有短暂的生命周期,可以被动态地创建、更新和删除。当Pod中的某个容器终止或失败时,Kubernetes会根据设置的重启策略自动重启Pod。这种自动化管理机制确保了应用程序的高可用性和可靠性。

例如,我们可以使用Pod来封装应用,将应用及其依赖项封装到单个Pod中,简化了应用的部署和管理。在微服务架构中,每个服务可以作为一个或多个Pod运行,以实现服务的独立部署和扩展。通过增加Pod的副本数,可以轻松实现应用的水平扩展,以应对流量增长。结合StatefulSet、DaemonSet等资源对象,Pod可以支持更复杂的部署模式,如状态管理服务和守护进程服务等。

总之,Pod在Kubernetes中扮演着重要的角色,可以帮助组织和管理容器化应用程序,并提供灵活的部署选项。

2. 如何使用Pod

在集群运维中,经常涉及Pod的创建、更新、删除和查看等操作,这里通过创建一个Pod的示例来加强对Pod的理解。通常,以下情况我们可能需要创建Pod。

- 应用程序需要多个容器协同工作:如果一个应用程序由多个相互依赖的容器组成,可以使用Pod将这些容器组合在一起,共享网络和存储资源。
- 需要对容器进行编排和管理:Pod提供了一种方便的方式来管理和编排容器,可以确保它们按照预期的方式运行。
- 需要实现负载均衡和服务发现:Pod可以通过Service暴露给外部访问,并自动实现负载均衡和服务发现。
- 需要扩展应用程序:通过创建多个副本的Pod,可以实现应用程序的水平扩展,以满足更高的负载需求。

- 需要隔离应用程序的不同部分：使用不同的Pod可以将应用程序的不同部分隔离开来，提高安全性和可维护性。
- 需要处理复杂的网络拓扑：Pod可以定义复杂的网络拓扑结构，例如多主机通信、跨集群通信等。

要创建一个Pod，可以使用Kubernetes的命令行工具Kubectl，例如：

```
kubectl run my-pod --image=my-image --restart=Never --port 8080 --expose=True
```

在这个命令中，run是Kubectl的一个子命令，用于创建新的资源。my-pod是用户给Pod起的名字，可以根据需要自定义。--image参数指定了容器镜像的名称，需要将其替换为实际的镜像名称。--restart参数用于设置Pod重启策略，这里设置为Never，表示不自动重启。--port参数暴露容器的端口，这里设置为8080。--expose参数会自动创建一个Service来暴露这个Pod。

执行上述命令后，Kubernetes会将Pod的配置信息存储到etcd中，并通知kube-scheduler选择一个合适的节点来部署Pod。一旦Pod被成功创建，可以通过以下命令查看Pod的状态：

```
kubectl get pods
```

这将显示所有正在运行的Pod及其状态。如果只想查看特定Pod的信息，可以使用以下命令：

```
kubectl describe pod my-pod
```

这将显示有关my-pod的详细信息，包括IP地址、容器状态等。

需要注意的是，这只是创建Pod的基本步骤。在实际生产环境中，可能还需要配置其他选项，如环境变量、卷挂载、资源限制等。这些可以通过在kubectl run命令中添加相应的参数来实现。

2.1.2 Kubectl 命令行工具

Kubectl是Kubernetes集群运维中重要的命令行工具，用于与Kubernetes集群进行交互和管理。

1. Kubectl的使用格式

Kubectl提供了多种子命令和参数，帮助用户执行各种操作，比如创建、更新、删除资源，以及查看集群状态等。其使用格式如下：

```
kubectl [command] [TYPE] [NAME] [flags]
```

参数说明：

- command部分：指定要对资源执行的操作，如create（创建）、get（获取）、describe（描述）和delete（删除）等。
- TYPE部分：指定资源类型。Kubernetes中的资源类型大小写敏感，并且可以以单数、复数或简写形式表示。例如：

```
kubectl get pod mypod
kubectl get pods mypod
kubectl get po mypod
```

以上三种形式都是等效的，均用于获取名为mypod的Pod信息。

- NAME部分：指定资源对象的名称，也是区分大小写的。如果省略名称，系统将返回属于该类型的全部资源对象列表。例如，kubectl get pods将返回所有Pod的列表。
- flags部分：这是Kubectl子命令的可选参数。例如，使用-s或--server参数指定Kubernetes API Server的URL地址而不使用默认值。

2. Kubectl命令示例

以下是一些常用的Kubectl命令示例。

（1）获取集群信息：

```
kubectl cluster-info
```

（2）获取所有命名空间中的Pod列表：

```
kubectl get pods --all-namespaces
```

（3）创建一个名为my-pod的Pod：

```
kubectl create deployment my-pod --image=nginx
```

（4）描述一个名为my-pod的Pod：

```
kubectl describe pod my-pod
```

（5）删除一个名为my-pod的Pod：

```
kubectl delete pod my-pod
```

（6）运行一个名为my-pod的Pod：

```
kubectl run my-pod --image=nginx
```

（7）将一个名为my-pod的Pod暴露为服务：

```
kubectl expose pod my-pod --port=80 --type=LoadBalancer
```

（8）获取所有服务的列表：

```
kubectl get services -A
```

（9）获取特定命名空间中的所有服务列表：

```
kubectl get services --namespace=my-namespace
```

（10）编辑一个名为my-service的服务：

```
kubectl edit service my-service
```

2.1.3 YAML 文件

1. 什么是YAML文件

Kubernetes使用YAML（或JSON）文件来声明和配置其资源需求。这些文件通常被命名为.yaml并存放在本地文件系统中，或者直接通过Kubectl命令行工具在命令中指示。

YAML文件的作用如下。

- 描述资源配置：YAML文件用于详细描述一个或多个Kubernetes对象（如Pods、Services、Deployments等）。
- 部署资源：通过Kubectl工具应用这些配置文件，可以在Kubernetes集群中创建或更新相应的资源。
- 环境抽象：YAML文件使得基础设施即代码（Infrastructure as Code，IaC）成为可能，便于版本控制和团队协作。

2. YAML文件格式

一个YAML文件由以下内容构成。

- 键-值对：使用冒号（:）分隔键和值，冒号后面通常需要跟一个空格，例如name: Dean。
- 列表：使用短横线（-）和缩进来表示列表项。例如：

```
- item1
- item2
- item3
```

- 字典（或称为映射）：通过键-值对来表示，使用缩进来表示层级关系。例如：

```
person:
  name: Dean
  age: 27
  children:
    - name: DeanBoys
      age: 5
    - name: DeanGirl
      age: 7
```

3. YAML文件示例

一个简单的Pod配置YAML文件可能看起来是这样的：

```
apiVersion: v1
kind: Pod
```

```yaml
metadata:
  name: myapp-pod
  labels:
    app: myapp
spec:
  containers:
  - name: myapp-container
    image: myapp:1.0.0
    ports:
    - containerPort: 8080
```

该文件定义了一个名为myapp-pod的Pod，包含一个容器，使用myapp:1.0.0镜像，并将容器的8080端口暴露出来。

参数说明：

- **apiVersion**：指定Kubernetes API的版本，影响资源的定义和行为。
- **kind**：指明资源的类型（例如Pod、Service等）。
- **metadata**：包含资源的元数据，如名称、标签等。
- **spec**：资源配置的详细规范，这通常是一个嵌套的对象，包含所有具体的配置信息。
- **containers**：在Pod资源配置中，这是容器配置的列表，每个列表项定义一个容器。
- **image**：容器使用的镜像。
- **name**：资源的名称，必须是唯一的。
- **labels**：键-值对，用于标识资源，便于管理和选择。

请注意，理解并正确使用YAML文件是进行有效的Kubernetes管理和开发的关键。

2.2 Replication

Replication Controller（RC，复制控制器）的主要作用是确保Pod的指定数量在Kubernetes集群中始终保持运行状态，并维护应用的高可用性和可靠性。它通过自动替换故障的Pod、动态调整Pod数量来适应负载变化，以及支持滚动升级和回滚功能来实现这一目标。

Replication控制器是早期Kubernetes的资源对象之一，目前在Kubernetes中已被ReplicaSet控制器替代，尽管如此，理解RC的基本概念和功能仍然对于学习和使用Kubernetes具有重要意义。本节通过一个实例来了解RC的具体使用。

本实例旨在通过编写资源清单来部署一个Replication控制器，并进一步实践对该控制器管理的Pod进行动态扩展（扩容）与缩减（缩容）操作，以增强对Kubernetes集群中Pod自动管理能力的理解与应用。

具体操作步骤如下：

01 在k8s-master节点创建命名空间：

```
[root@k8s-master ~]# kubectl create ns dean
```

参数解析：

- kubectl create：该命令通常与YAML或JSON格式的文件一起使用，这些文件描述了要创建的资源及其属性。不过，kubectl create也支持一些简单资源（如Pod）的命令行参数创建。
- ns：Namespace（命名空间）是一个核心概念，它允许用户将集群中的资源划分为多个虚拟的、逻辑上隔离的子集。

02 在k8s-master节点创建资源清单文件：

```
[root@k8s-master ~]# vim rc-nginx.yaml
---
# 控制器的类型
kind: ReplicationController
# API版本
apiVersion: v1
# 元数据信息
metadata:
  # 资源的名字
  name: rc-nginx-pod
  # 资源所属命名空间
  namespace: dean
spec:
  # 副本数
  replicas: 3
  # 用于选择具有特定标签（Labels）的资源对象的机制
  selector:
    app: rc-nginx
  template:
    metadata:
      name: rc-nginx-container
      namespace: dean
      labels:
        app: rc-nginx
    spec:
      containers:
        # 容器名
        - name: rc-nginx-container
          # 镜像信息
          image: nginx:1.19
          # 镜像拉取策略
          # Always: 每次启动容器时，都会从镜像仓库中拉取最新的镜像。这是默认的镜像拉取策略，但需
要注意的是，在配置文件中明确指定了imagePullPolicy为IfNotPresent或Never时，才会覆盖这一默认行为
```

```
                # IfNotPresent: 如果节点上已经存在相同的镜像，则直接使用节点上的镜像；如果节点上没有该
镜像，则从镜像仓库中拉取最新的镜像。这种策略有助于减少不必要的网络流量和镜像仓库负载，同时确保节点上已缓
存的镜像得到利用
                # Never: 只使用节点上已经存在的镜像，不会从镜像仓库中拉取镜像。如果节点上没有所需的镜像，
则Pod启动失败，并报告错误
                imagePullPolicy: IfNotPresent
                ports:
                  - containerPort: 80
```

03 在k8s-master节点应用资源清单：

```
[root@k8s-master ~]#kubectl apply -f rc-nginx.yaml
replicationcontroller/rc-nginx-pod created
```

使用Kubectl工具将rc-nginx.yaml文件中定义的资源应用到Kubernetes集群中。使用-f选项指定要应用的文件名。

04 查看Pod状态信息：

```
[root@k8s-master ~]# kubectl get pods -n dean
NAME                   READY   STATUS    RESTARTS   AGE
rc-nginx-pod-8hx29     1/1     Running   0          109s
rc-nginx-pod-w6kk6     1/1     Running   0          109s
rc-nginx-pod-w9nb5     1/1     Running   0          109s
```

05 扩容Pod副本：

```
[root@k8s-master ~]# kubectl scale rc --replicas=6 -n dean rc-nginx-pod
replicationcontroller/rc-nginx-pod scaled
```

参数解析：

- kubectl scale：这是主命令，用于改变资源的副本数。
- rc：这是资源类型，表示ReplicationController。但在现代Kubernetes集群中，ReplicationController已经逐渐被更高级的控制器（如Deployment）所取代，因为Deployment提供了更多的功能，如滚动更新、回滚等。
- --replicas=6：这是一个参数，指定将资源的副本数设置为6。
- -n dean：这是一个命名空间标志（flag），用于指定命令应在哪个命名空间中执行。在这里，命令将在名为dean的命名空间中执行。
- rc-nginx-pod：这是要扩容的ReplicationController的名称。

06 查看Pod状态信息：

```
--- 扩容的Pod正在创建中...
[root@k8s-master ~]# kubectl get pods -n dean
NAME                   READY   STATUS              RESTARTS   AGE
rc-nginx-pod-79sdf     0/1     ContainerCreating   0          5s
rc-nginx-pod-8hx29     1/1     Running             0          3m47s
```

```
rc-nginx-pod-ffzd4      0/1     ContainerCreating   0       5s
rc-nginx-pod-sn7mn      0/1     ContainerCreating   0       5s
rc-nginx-pod-w6kk6      1/1     Running             0       3m47s
rc-nginx-pod-w9nb5      1/1     Running             0       3m47s
--- 等待几秒后,再次查看状态为Running
[root@k8s-master ~]# kubectl get pods -n dean
NAME                    READY   STATUS      RESTARTS    AGE
rc-nginx-pod-79sdf      1/1     Running     0           29s
rc-nginx-pod-8hx29      1/1     Running     0           4m11s
rc-nginx-pod-ffzd4      1/1     Running     0           29s
rc-nginx-pod-sn7mn      1/1     Running     0           29s
rc-nginx-pod-w6kk6      1/1     Running     0           4m11s
rc-nginx-pod-w9nb5      1/1     Running     0           4m11s
```

07 缩容Pod副本:

```
[root@k8s-master ~]# kubectl scale rc --replicas=3 -n dean rc-nginx-pod
replicationcontroller/rc-nginx-pod scaled
```

这个命令使用Kubectl工具将名为dean的命名空间中名为rc-nginx-pod的ReplicationController的副本数量调整为3。

08 查看Pod状态信息:

```
--- 扩容的Pod已经被删除
[root@k8s-master ~]# kubectl get pods -n dean
NAME                    READY   STATUS      RESTARTS    AGE
rc-nginx-pod-8hx29      1/1     Running     0           10m
rc-nginx-pod-w6kk6      1/1     Running     0           10m
rc-nginx-pod-w9nb5      1/1     Running     0           10m
```

本实例首先在k8s-master节点创建命名空间,然后创建资源清单文件rc-nginx.yaml,其中定义了控制器的类型、副本数、选择器和模板等信息。接着应用资源清单文件到Kubernetes集群中,成功创建了名为rc-nginx-pod的Replication控制器。通过查看Pod状态信息,可以看到有三个副本正在运行。

为了测试扩容功能,使用kubectl scale命令将副本数增加到6,再次查看Pod状态信息,可以看到新的Pod正在创建中。等待一段时间后,再次查看状态为Running,确认扩容成功。

最后,为了测试缩容功能,使用kubectl scale命令将副本数减少到3,再次查看Pod状态信息,可以看到多余的Pod已经被删除。

2.3 ReplicaSet

ReplicaSet控制器(简称RS)用于确保Pod的指定数量在集群中始终保持运行状态。与早期的Replication Controller(RC)相比,ReplicaSet提供了更加灵活和强大的功能,特别是在标签选择方面。在新版本的Kubernetes中,推荐使用ReplicaSet来取代Replication Controller。

ReplicaSet的主要作用如下。

- 确保Pod数量：ReplicaSet负责维持集群中Pod的副本数量，如果Pod因故障停止运行，ReplicaSet会自动创建新的Pod以替换故障的Pod，确保Pod总数保持不变。
- 增强的标签选择：ReplicaSet支持使用集合式的选择器，这使得Pod的选择和管理更加灵活和精确。例如，可以基于特定的应用版本或功能分组来管理Pod。
- 滚动更新和应用管理：ReplicaSet还支持滚动更新，允许开发者逐步替换旧的Pod版本，从而无缝地升级应用，减少更新过程中的服务中断时间。
- 动态扩缩容：根据工作负载的变化，ReplicaSet可以动态调整Pod的数量，优化资源使用效率。
- 简化的应用部署：通过使用ReplicaSet，开发者可以更加简便地部署和管理应用，无须手动干预Pod的创建和删除。
- 环境适应性：ReplicaSet允许通过定义不同的模板和选择器来适应开发、测试和生产等多种环境的需求。
- 高可用性和自恢复：自动替换故障的Pod，提高了系统的可靠性和可用性。
- 标签和选择器的高级用法：ReplicaSet支持更复杂的标签和选择器，使得对Pod的管理更加精细化。

此外，ReplicaSet还支持键-值对的选择形式，并支持matchExpressions字段，可以提供多种选择。目前支持的操作如下。

- In：匹配具有指定值的标签。
- NotIn：匹配不具有指定值的标签。
- Exists：匹配存在指定标签的Pod。
- DoesNotExist：匹配不存在指定标签的Pod。

ReplicaSet不仅确保了Pod的指定数量，还通过其高级的标签选择功能，提供了更加强大和灵活的Pod管理机制。这些特性使得ReplicaSet成为现代Kubernetes环境中不可或缺的资源对象。

接下来通过一个实例来演示ReplicaSet的具体使用。

本实例旨在通过YAML格式定义ReplicaSet的配置，实现Pod的自动部署与副本管理。首先，修改特定Pod的标签，展示如何在不直接修改ReplicaSet定义的情况下，直接修改其管理下的某个Pod的标签，理解Pod标签与ReplicaSet选择器之间的关系及其影响。然后运用kubectl命令的格式化输出选项以更清晰、更易读的格式查看Kubernetes的资源状态。接下来使用Kubectl进行资源删除时，可以启用级联删除功能（默认启用），确保删除操作不仅移除目标资源本身，还自动清理与之关联的所有子资源。最后实践如何在必要时禁用级联删除，仅删除指定资源而不影响其依赖的子资源。

具体操作步骤如下:

01 在k8s-master节点创建命名空间:

```
[root@k8s-master ~]# kubectl create ns dean
```

02 在k8s-master节点创建资源清单文件:

```
[root@k8s-master ~]# vim rc-nginx.yaml
apiVersion: apps/v1
kind: ReplicaSet
metadata:
  name: rs-nginx-pod
  namespace: dean
spec:
  replicas: 3
  selector:
    matchLabels:
      tier: rs-nginx
  template:
    metadata:
      labels:
        tier: rs-nginx
    spec:
      containers:
        - name: rs-nginx-container
          image: nginx:1.19
          imagePullPolicy: IfNotPresent
          env:
            - name: GET_HOSTS_FROM
              value: dns
```

03 应用资源:

```
[root@k8s-master ~]# kubectl apply -f rs-nginx.yaml
replicaset.apps/rs-nginx-pod created
```

04 使用Kubectl工具将rs-nginx.yaml文件中定义的资源应用到Kubernetes集群中。使用-f选项指定要应用的文件名。

05 查看Pod状态信息:

```
--- Pod正在创建中...
[root@k8s-master ~]# kubectl get pods -n dean
NAME                    READY   STATUS              RESTARTS   AGE
rs-nginx-pod-jskg4      0/1     ContainerCreating   0          4s
rs-nginx-pod-lxlcv      0/1     ContainerCreating   0          4s
rs-nginx-pod-vnpv8      0/1     ContainerCreating   0          4s
--- 查看Pod状态和标签
[root@k8s-master ~]# kubectl get pods -n dean --show-labels
NAME                 READY   STATUS   RESTARTS   AGE   LABELS
```

```
rs-nginx-pod-jskg4    1/1    Running    0    4m56s    tier=rs-nginx
rs-nginx-pod-lxlcv    1/1    Running    0    4m56s    tier=rs-nginx
rs-nginx-pod-vnpv8    1/1    Running    0    4m56s    tier=rs-nginx
```

06 更改一个Pod的标签为tier=rs-nginx-1：

```
[root@k8s-master ~]# kubectl label -n dean pods rs-nginx-pod-jskg4 tier=rs-nginx-1 --overwrite
pod/rs-nginx-pod-jskg4 labeled
```

07 查看Pod状态和标签：

```
[root@k8s-master ~]# kubectl get pods -n dean --show-labels
NAME                  READY  STATUS     RESTARTS  AGE   LABELS
rs-nginx-pod-jpncm    1/1    Running    0         23s   tier=rs-nginx
rs-nginx-pod-jskg4    1/1    Running    0         18m   tier=rs-nginx-1
rs-nginx-pod-lxlcv    1/1    Running    0         18m   tier=rs-nginx
rs-nginx-pod-vnpv8    1/1    Running    0         18m   tier=rs-nginx
```

参数解析：

- **kubectl**：是Kubernetes的命令行工具，用于与Kubernetes集群进行交互。
- **get pods**：指定要获取的资源类型为Pod。
- **-n dean或--namespace=dean**：指定操作所在的命名空间为dean。如果不指定命名空间，默认使用default命名空间。
- **--show-labels**：指示Kubectl在输出中包含每个Pod的标签。标签是附加到Kubernetes对象（如Pod、Deployment、Service等）上的键-值对，用于组织和选择一组对象。

Pod的列表信息解析如下。

- 现象：我们发现有4个一样的Pod（rs-nginx-x），其中有一个Pod的标签是更改后的标签（tier=rs-nginx-1）。
- 原因：我们将一个Pod的标签更改后，这个Pod就不属于RS控制下的，RS判断属于它的标签只存在两个，根据资源清单中的副本数要求是3，所以会立即创建一个Pod，标签为tier=rs-nginx-1。

08 K8s格式化输出：

```
[root@k8s-master ~]# kubectl get pods -n dean -l tier=rs-nginx -o custom-columns=Name:metadata.name,Image:spec.containers[0].image
# 输出信息
Name                  Image
rs-nginx-pod-jpncm    nginx:1.19
rs-nginx-pod-jskg4    nginx:1.19
rs-nginx-pod-lxlcv    nginx:1.19
rs-nginx-pod-vnpv8    nginx:1.19
```

该命令列出了所有在dean命名空间中，标签为tier=rs-nginx的Pod的名称和它们第一个容器的镜像。这样的输出格式非常有用，尤其是在想要快速查看具有特定标签的Pod及其镜像时。

参数解析：

- kubectl get pods：这是使用Kubectl工具获取Pod资源的命令。
- -n dean：指定操作的命名空间为dean。
- -l tier=rs-nginx：使用标签选择器（label selector）来过滤资源。这里选择了所有标签中包含tier=rs-nginx的Pod。
- -o custom-columns=Name:metadata.name,Image:spec.containers[0].image：指定输出的自定义列。这里指定了两列，一列是Pod的名称（从metadata.name字段获取），另一列是Pod中第一个容器的镜像（从spec.containers[0].image字段获取）。注意，这个命令假设每个Pod至少有一个容器。

09 级联删除/非级联删除：

（1）级联删除：控制器和Pod都会被删除。

（2）非级联删除：控制器被删除，Pod会变成孤儿，删除后不会重建。

```
Ps：参数：--cascade=false已弃用，最新使用参数：--cascade=orphan
[root@k8s-master ~]# kubectl delete rs rs-nginx-pod --cascade=orphan -n dean
replicaset.apps "rs-nginx-pod" deleted
```

这里使用Kubectl命令行工具删除一个ReplicaSet资源，使用--cascade=orphan参数保留ReplicaSet资源下的Pod。

非级联删除后，查看控制器和Pod的状态：

```
--- Pod成为孤儿
[root@k8s-master ~]# kubectl get pods -n dean
NAME                   READY   STATUS    RESTARTS   AGE
rs-nginx-pod-8s8kk     1/1     Running   0          91m
rs-nginx-pod-jskg4     1/1     Running   0          4h57m
rs-nginx-pod-kjzmh     1/1     Running   0          91m
rs-nginx-pod-nglfs     1/1     Running   0          91m

--- 发现RS控制器下不存在资源
[root@k8s-master ~]# kubectl get rs -n dean
No resources found in dean namespace.
```

10 删除资源清单，会删除该资源清单下的Pod：

```
[root@k8s-master ~]# kubectl delete -f rs-nginx.yaml
replicaset.apps "rs-nginx-pod" deleted
```

用于Kubectl工具删除在Kubernetes集群中由rs-nginx.yaml文件定义的资源的命令。

⑪ 查看Pod信息，只剩下tier=rs-nginx-1标签的Pod，因为该Pod不受RS管理：

```
[root@k8s-master ~]# kubectl get pods -n dean --show-labels
NAME                   READY   STATUS    RESTARTS   AGE   LABELS
rs-nginx-pod-jskg4     1/1     Running   0          55m   tier=rs-nginx-1
```

2.4 Deploymant

Deployment（简称Deploy）是Kubernetes控制器的又一种实现形式，它基于ReplicaSet控制器构建，可为Pod和ReplicaSet资源提供声明式更新。相比之下，Pod和ReplicaSet属于较低级别的资源，它们通常不会直接被使用。

2.4.1 Deployment 概述

Deployment控制器的主要任务是确保Pod资源的稳定运行，其大部分功能都能通过调用ReplicaSet控制器来实现，同时还增加了以下特性。

（1）事件和状态查看：在需要时，可以查看Deployment对象升级的详细进度和状态。

（2）回滚：支持使用回滚机制将应用恢复到前一个或用户指定的历史版本。

（3）版本记录：保存对Deployment对象的每次操作，以便进行回滚。

（4）暂停和启动：对于每一次升级，都能够随时暂停和启动。

（5）多种自动更新方案：

- Recreate，即重建更新机制，全面停止、删除旧有的Pod后用新版本替代。
- RollingUpdate，即滚动升级机制，逐步替换旧有的Pod至新的版本。

2.4.2 Deployment 创建与访问

可以通过以下命令行直接创建一个Deployment服务：

```
[root@k8s-master ~]# kubectl create deployment mynginx --image=nginx:1.19 -n dean
deployment.apps/mynginx created
```

该命令是在Kubernetes集群的dean命名空间中创建一个名为mynginx的Deployment，该Deployment会管理运行nginx:1.19镜像的Pod。通过这种方式，很容易在Kubernetes集群中部署和管理Nginx容器实例。

查看Deployment控制器下的资源：

```
[root@k8s-master ~]# kubectl get deploy -n dean
NAME      READY   UP-TO-DATE   AVAILABLE   AGE
mynginx   1/1     1            1           12s
```

查看Pod资源信息：

```
[root@k8s-master ~]# kubectl get pods -n dean
NAME                        READY   STATUS    RESTARTS   AGE
mynginx-664b8b966c-c24wb    1/1     Running   0          20s
rs-nginx-pod-8s8kk          1/1     Running   0          21h
rs-nginx-pod-jskg4          1/1     Running   0          24h
rs-nginx-pod-kjzmh          1/1     Running   0          21h
rs-nginx-pod-nglfs          1/1     Running   0          21h
```

通过kubectl expose创建svc：

```
[root@k8s-master ~]# kubectl expose deployment mynginx --name=mynginx --port=80 --target-port=80 --type=NodePort -n dean
service/mynginx exposed
```

该命令是在Kubernetes集群的dean命名空间中，基于名为mynginx的Deployment创建一个新的NodePort类型的Service，该Service的名称为mynginx，它监听集群内部的80端口，并将流量转发到Pod的80端口上。由于Service是NodePort类型的，因此它会将流量映射到集群中每个节点的某个静态端口上（这个端口是Kubernetes自动分配的，并且可以通过kubectl get svc mynginx -n dean命令查看），从而允许用户从集群外部访问Nginx容器。

参数解析：

- kubectl expose：这是Kubectl命令行工具的一个子命令，用于基于现有的资源（如Pod、Deployment等）创建新的Service资源。
- deployment mynginx：这指定了要基于哪个Deployment创建Service。在这个例子中，基于名为mynginx的Deployment。
- --name=mynginx：这指定了新创建的Service的名称。在这个例子中，Service的名称也是mynginx，但通常建议为Service选择一个描述性更强或与服务功能更相关的名称，以避免与Deployment名称混淆。不过，这里使用相同的名称是可行的。
- --port=80：这指定了Service的端口号，即集群内部访问Service时使用的端口。在这个例子中，它设置为80，这是HTTP服务的标准端口。
- --target-port=80：这指定了Service将流量转发到Pod上哪个端口。在这个例子中，它也设置为80，意味着Service会将流量转发到Pod上运行的Nginx容器的80端口。
- --type=NodePort：这指定了Service的类型。在Kubernetes中，Service有多种类型，如ClusterIP、NodePort、LoadBalancer等。NodePort类型会将Service映射到集群中每个节点的静态端口上，从而使Service可以从集群外部访问。在这个例子中，Service被设置为NodePort类型。
- -n dean或--namespace=dean：这指定了Service应该在哪个命名空间中创建。在这个例子中，Service被创建在名为dean的命名空间中。

查看Service信息并访问服务：

```
--- 查看dean命名空间下的service信息
[root@k8s-master ~]# kubectl get svc -n dean
NAME       TYPE       CLUSTER-IP       EXTERNAL-IP    PORT(S)         AGE
mynginx    NodePort   10.100.179.145   <none>         80:30945/TCP    4s

--- curl访问服务
[root@k8s-master ~]# curl 192.168.1.10:30945
<!DOCTYPE html>
<html>
<head>
<title>Welcome to nginx!</title>
<style>
    body {
        width: 35em;
        margin: 0 auto;
        font-family: Tahoma, Verdana, Arial, sans-serif;
    }
</style>
</head>
<body>
<h1>Welcome to nginx!</h1>
<p>If you see this page, the nginx web server is successfully installed and
working. Further configuration is required.</p>

<p>For online documentation and support please refer to
<a href="http://nginx.org/">nginx.org</a>.<br/>
Commercial support is available at
<a href="http://nginx.com/">nginx.com</a>.</p>

<p><em>Thank you for using nginx.</em></p>
</body>
</html>
```

2.4.3 滚动更新策略实例

本实例的目标是利用kubectl set image命令高效地更新Kubernetes中部署的应用镜像版本，以实现应用的快速迭代与Recreate策略的应用。通过这一命令，我们能够直接指定新的镜像版本，并触发Pod的重建过程，确保应用服务无缝地迁移到新的镜像版本上。这样的操作不仅简化了镜像版本更新的流程，还增强了部署的灵活性和效率。

在这个滚动更新策略实例中，涉及以下滚动更新策略，读者可以在YAML文件中根据需要更改这些参数。

- spec.strategy.type：更新Deployment的方式，默认是RollingUpdate。
- RollingUpdate：滚动更新，可以指定maxSurge和maxUnavailable。

- maxUnavailable：指定在回滚或更新时最大不可用的Pod的数量，默认为25%，可以设置成数字或百分比，如果该值为0，那么maxSurge不能为0。
- maxSurge：可以超过期望值的最大Pod数，可选字段，默认为25%，可以设置成数字或百分比，如果该值为0，那么maxUnavailable不能为0。
- Recreate：重建，先删除旧的Pod，再创建新的Pod。

涉及的其他参数：

- .spec.revisionHistoryLimit：设置保留RS旧的revision的个数，若设置为0，则不保留历史数据。
- .spec.minReadySeconds：可选参数，指定新创建的Pod在没有任何容器崩溃的情况下视为Ready最小的秒数，默认为0，即一旦被创建就视为可用。

下面介绍滚动更新策略实例的具体操作步骤。

01 在k8s-master节点创建资源清单：

```
[root@k8s-master ~]# vim nginx-Recreate.yaml
apiVersion: apps/v1
kind: Deployment
metadata:
  name: nginx-recreate
  namespace: dean
spec:
  revisionHistoryLimit: 10000
  minReadySeconds: 3
  strategy:
    type: Recreate
  selector:
    matchLabels:
      app: nginx-recreate
  replicas: 6
  template:
    metadata:
      labels:
        app: nginx-recreate
    spec:
      containers:
        - name: nginx
          image: nginx:1.19
          imagePullPolicy: IfNotPresent
          ports:
            - containerPort: 80
```

在k8s-master节点上创建名为nginx-Recreate的资源清单文件nginx-Recreate.yaml，其中定义了一个使用Recreate策略的Deployment，包含6个副本和Nginx容器。

02 应用资源：

```
[root@k8s-master ~]# kubectl apply -f nginx-Recreate.yaml
```

03 查看Pod状态：

```
[root@k8s-master ~]# kubectl get pods -n dean
NAME                             READY   STATUS    RESTARTS   AGE
nginx-recreate-7f6bdd94fd-28kkh  1/1     Running   0          8m53s
nginx-recreate-7f6bdd94fd-8gc7t  1/1     Running   0          8m53s
nginx-recreate-7f6bdd94fd-br4qt  1/1     Running   0          8m53s
nginx-recreate-7f6bdd94fd-tgzk4  1/1     Running   0          8m53s
nginx-recreate-7f6bdd94fd-vv5wv  1/1     Running   0          8m53s
nginx-recreate-7f6bdd94fd-xls2c  1/1     Running   0          8m53s
```

04 通过kubectl set image命令更改镜像版本：

```
[root@k8s-master ~]# kubectl set image deployment/nginx-recreate nginx=nginx:1.20 -n dean
```

该命令是在Kubernetes集群的dean命名空间中，将名为nginx-recreate的Deployment中名为nginx的容器的镜像更新为nginx:1.20。

参数解析：

- kubectl set image：这是Kubectl命令行工具的一个子命令，用于更新Kubernetes集群中资源的镜像版本。
- deployment/nginx-recreate：这指定了要更新镜像的Deployment的名称和类型。在这个例子中，它是名为nginx-recreate的Deployment。
- nginx=nginx:1.20：这指定了要更新的容器名称和新的镜像版本。在这个例子中，它会找到Deployment中名为nginx的容器，并将其镜像更新为nginx:1.20。这意味着所有使用该Deployment创建的Pod都将使用新的Nginx镜像版本。
- -n dean或--namespace=dean：这指定了命令应该在哪个命名空间下执行。在这个例子中，命令在dean命名空间中执行。

05 查看Pod重建状态：

```
--- 正在删除旧Pod
[root@k8s-master ~]# kubectl get pods -n dean
NAME                             READY   STATUS        RESTARTS   AGE
nginx-recreate-7f6bdd94fd-28kkh  0/1     Terminating   0          17m
nginx-recreate-7f6bdd94fd-8gc7t  1/1     Terminating   0          17m
nginx-recreate-7f6bdd94fd-br4qt  1/1     Terminating   0          17m
nginx-recreate-7f6bdd94fd-tgzk4  1/1     Terminating   0          17m
nginx-recreate-7f6bdd94fd-vv5wv  1/1     Terminating   0          17m
nginx-recreate-7f6bdd94fd-xls2c  0/1     Terminating   0          17m
--- 创建新的Pod
```

```
[root@k8s-master ~]# kubectl get pods -n dean
NAME                                  READY   STATUS              RESTARTS   AGE
nginx-recreate-74d85b5c4b-29gm8       0/1     ContainerCreating   0          1s
nginx-recreate-74d85b5c4b-8665d       0/1     ContainerCreating   0          0s
nginx-recreate-74d85b5c4b-fg5nh       0/1     ContainerCreating   0          1s
nginx-recreate-74d85b5c4b-hd2tn       0/1     ContainerCreating   0          0s
nginx-recreate-74d85b5c4b-pz8dl       0/1     ContainerCreating   0          1s
nginx-recreate-74d85b5c4b-qtlhr       0/1     ContainerCreating   0          0s
--- Pod创建成功
[root@k8s-master ~]# kubectl get pods -n dean
NAME                                  READY   STATUS    RESTARTS   AGE
nginx-recreate-74d85b5c4b-29gm8       1/1     Running   0          68s
nginx-recreate-74d85b5c4b-8665d       1/1     Running   0          67s
nginx-recreate-74d85b5c4b-fg5nh       1/1     Running   0          68s
nginx-recreate-74d85b5c4b-hd2tn       1/1     Running   0          67s
nginx-recreate-74d85b5c4b-pz8dl       1/1     Running   0          68s
nginx-recreate-74d85b5c4b-qtlhr       1/1     Running   0          67s
```

观察Pod重建过程中的状态变化，包括正在删除旧Pod和创建新Pod的过程。同时，查看更新后的Pod状态，确认所有Pod都已成功运行并使用新的镜像版本。

本实例展示了如何在Kubernetes集群中使用滚动更新策略来更新Deployment中的容器镜像。

2.4.4 更新/回滚/暂停/恢复

本实例用于测试Kubernetes的部署历史记录功能（通过--record选项）、部署暂停与恢复能力（利用kubectl rollout pause与kubectl rollout resume命令）以及部署回滚机制（kubectl rollout undo命令）。这些实践将帮助你充分理解Kubernetes在部署管理上的强大能力，包括如何追踪变更历史、如何安全地暂停和恢复部署过程，以及在必要时如何快速回滚到之前的稳定版本。

在k8s-master节点创建资源清单：

```
[root@k8s-master ~]# vim nginx-Record.yaml
apiVersion: apps/v1
kind: Deployment
metadata:
  name: nginx-record
  namspace: dean
spec:
  selector:
    matchLabels:
      app: nginx-record
  replicas: 3
  template:
    metadata:
      labels:
        app: nginx-record
    spec:
      containers:
```

```
      - name: nginx
        image: nginx:1.19
        imagePullPolicy: IfNotPresent
        ports:
          - containerPort: 80
```

创建服务：

```
[root@k8s-master ~]# kubectl apply -f nginx-Record.yaml --record
Flag --record has been deprecated, --record will be removed in the future
deployment.apps/nginx-record created
```

该命令将nginx-Record.yaml文件中定义的资源应用到Kubernetes集群中，并记录下这次操作的详细信息以便后续追踪和审计，--record参数可以记录命令，我们可以很方便的查看每次revision的变化。

下面通过kubectl set image更改镜像版本。

格式一：

```
kubectl set image 控制器类型/控制器名 容器名=镜像名:镜像版本
```

格式二：

```
kubectl set image 控制器类型 控制器名 容器名=镜像名:镜像版本
```

在Kubernetes集群的dean命名空间中，将名为nginx-record的Deployment中名为nginx的容器的镜像更新为nginx:1.20：

```
[root@k8s-master ~]# kubectl set image deployment/nginx-record nginx=nginx:1.20 -n dean
```

查看更新版本历史：

```
[root@k8s-master ~]# kubectl rollout history -n dean deployment nginx-record
deployment.apps/nginx-record
REVISION  CHANGE-CAUSE
1         kubectl apply --filename=nginx-Record.yaml --record=true
2         kubectl apply --filename=nginx-Record.yaml --record=true
```

该命令会列出名为nginx-record的Deployment的所有历史版本，包括每个版本的创建时间、镜像信息以及可能的其他关键更改。

参数解析：

- Kubectl：Kubernetes的命令行工具，用于与Kubernetes集群进行交互。
- rollout history：Kubectl的一个子命令，用于查看Deployment的滚动更新历史。这个命令特别有用，因为它可以显示Deployment在不同时间点的状态，包括哪些更改导致了新的Pod部署。

- -n dean或--namespace=dean：指定命令应该在哪个命名空间下执行。在这个例子中，命令将在dean命名空间下执行。
- deployment nginx-record：指定要查看其历史版本的Deployment的名称。在这个例子中，查找名为nginx-record的Deployment。

查看Nginx版本，确认是否被更新：

```
--- 查看Pod信息
[root@k8s-master ~]# kubectl get pods -n dean
NAME                             READY   STATUS    RESTARTS   AGE
nginx-record-9dbc59d6b-g2lpp     1/1     Running   0          92s
nginx-record-9dbc59d6b-hnwgw     1/1     Running   0          96s
nginx-record-9dbc59d6b-l6ktf     1/1     Running   0          100s
--- 查看其中一个Pod的Nginx版本
[root@k8s-master ~]# kubectl exec -it -n dean nginx-record-9dbc59d6b-g2lpp -- nginx -v
nginx version: nginx/1.20.2
```

回滚到上一个版本，因为只有一个以前的版本，所以不使用revision参数也行：

```
[root@k8s-master ~]# kubectl rollout undo -n dean deployment nginx-record
deployment.apps/nginx-record rolled back
```

该命令将在Kubernetes集群中针对特定命名空间（dean）下的名为nginx-record的Deployment执行回滚操作。这意味着该命令会将名为nginx-record的Deployment恢复到它之前的一个版本。

参数解析：

- Kubectl：Kubernetes的命令行工具，用于与Kubernetes集群进行交互。
- rollout undo：Kubectl的一个子命令，用于回滚Deployment到之前的版本。
- -n dean或--namespace=dean：指定命令应该在哪个命名空间下执行。在这个例子中，命令将在dean命名空间下执行。
- deployment nginx-record：指定要执行回滚操作的Deployment的名称。在这个例子中，查找名为nginx-record的Deployment。

查看Pod信息：

```
--- 正在创建新的Pod
[root@k8s-master ~]# kubectl get pods -n dean
NAME                             READY   STATUS              RESTARTS   AGE
nginx-record-95b8bf58-sdwrq      0/1     ContainerCreating   0          3s
nginx-record-9dbc59d6b-g2lpp     1/1     Running             0          8m10s
nginx-record-9dbc59d6b-hnwgw     1/1     Running             0          8m14s
nginx-record-9dbc59d6b-l6ktf     1/1     Running             0          8m18s
--- 删除旧的Pod
[root@k8s-master ~]# kubectl get pods -n dean
NAME                             READY   STATUS              RESTARTS   AGE
```

```
nginx-record-95b8bf58-2jf99      0/1    Pending        0    0s
nginx-record-95b8bf58-sdwrq      1/1    Running        0    5s
nginx-record-9dbc59d6b-g2lpp     1/1    Running        0    8m12s
nginx-record-9dbc59d6b-hnwgw     1/1    Running        0    8m16s
nginx-record-9dbc59d6b-l6ktf     1/1    Terminating    0    8m20s
--- 等待片刻，可以看到旧的 Pod 已被删除完
[root@k8s-master ~]# kubectl get pods -n dean
NAME                             READY  STATUS         RESTARTS  AGE
nginx-record-95b8bf58-2jf99      1/1    Running        0         43s
nginx-record-95b8bf58-cspkd      1/1    Running        0         39s
nginx-record-95b8bf58-sdwrq      1/1    Running        0         48s
```

查看 RS 控制器：

```
--- 已经回滚到旧版本
[root@k8s-master ~]# kubectl get rs -n dean
NAME                      DESIRED  CURRENT  READY  AGE
nginx-record-95b8bf58     3        3        3      34m
nginx-record-9dbc59d6b    0        0        0      16m
```

查看 Nginx 版本，确认是否被回滚：

```
[root@k8s-master ~]# kubectl exec -it -n dean nginx-record-95b8bf58-2jf99 -- nginx -v
nginx version: nginx/1.19.10
```

该命令会连接到 dean 命名空间下的 nginx-record-95b8bf58-2jf99 的 Pod，并在其中的主容器（默认情况下）中执行 nginx -v 命令，以显示 Nginx 的版本信息。如果 Pod 有多个容器，则可能需要使用 -c 选项来指定要执行命令的容器。

参数解析：

- kubectl exec：Kubectl 的一个子命令，用于在 Kubernetes Pod 的容器内执行命令。
- -it：这是两个选项的组合。-i 或 --interactive 保持 STDIN 打开，即使没有附加到终端；-t 或 --tty 分配一个伪终端。这两个选项一起使用时，允许与 Pod 中的容器进行交互式会话。
- -n dean：指定命令应该在哪个命名空间（dean）下执行。
- nginx-record-95b8bf58-2jf99：这是要执行命令的 Pod 的名称。然而，这个名称看起来像是一个自动生成的 Pod 名称，通常基于 Deployment 或 StatefulSet 等控制器的名称加上一个哈希或随机数来确保唯一性。这意味着如果想要针对一个由 Deployment 管理的 Pod 执行命令，通常需要使用更通用的标签选择器或 Pod 模板中的名称（尽管后者可能不够具体，因为会有多个 Pod 实例）。
- --：一个常见的命令行约定，用于明确区分命令的选项（或称为 flags）和传递给命令本身的参数。然而，在 kubectl exec 的上下文中，-- 并不是严格必需的，因为 kubectl exec 命令通常不需要区分其选项和传递给目标 Pod 容器内命令的参数。
- nginx -v：查看 Nginx 版本的命令。

查看回滚状态，最后显示successfully rolled out，表示已经回滚成功：

```
[root@k8s-master ~]# kubectl rollout status -n dean deployment nginx-record
deployment "nginx-record" successfully rolled out
```

该命令用于在Kubernetes集群中查询特定命名空间（dean）下名为nginx-record的Deployment的滚动更新（rollout）状态。这个命令会等待直到Deployment的滚动更新完成，或者直到它检测到某些问题（如Deployment失败或超时）。

暂停Deployment的更新，暂停后不可以进行操作回滚：

```
[root@k8s-master ~]# kubectl rollout pause -n dean deployment nginx-record
deployment.apps/nginx-record paused
```

该命令用于在Kubernetes集群中暂停特定命名空间（dean）下名为nginx-record的Deployment的新版本的滚动更新。这意味着，即使之后更新了Deployment的配置（比如改变了镜像版本或环境变量），这些更新也不会被应用到Pod上，因为滚动更新过程被暂停了，设置暂停后，以后再使用undo回滚命令就会报错。

恢复暂停的更新，恢复了才可以进行操作回滚：

```
[root@k8s-master ~]# kubectl rollout resume -n dean deployment nginx-record
deployment.apps/nginx-record resumed
```

该命令用于在Kubernetes集群中恢复特定命名空间（dean）下名为nginx-record的Deployment的滚动更新。在使用kubectl rollout pause命令暂停了Deployment的滚动更新后，可以使用kubectl rollout resume命令继续更新过程。

2.4.5 扩容的3种方式

扩容是指增加计算资源以满足应用程序或服务的需求。在Kubernetes中，扩容通常指的是增加Pod的数量，以便更好地处理负载和提高系统的可用性。

当应用程序或服务的负载增加时，现有的Pod可能无法满足需求，导致性能下降或响应时间延长。通过扩容可以增加更多的Pod来分担负载，从而提高系统的吞吐量和响应能力。此外，扩容还可以提供更好的容错能力，因为即使某个Pod发生故障，其他Pod仍然可以继续提供服务。

我们可以通过以下三种方式来达到扩容的目的。

（1）使用命令行工具Kubectl进行扩容：

```
[root@k8s-master ~]# kubectl scale deployment nginx-record --replicas 10
```

该命令将名为nginx-record的Deployment的副本数（replicas）设置为10。

（2）直接修改YAML文件的方式进行扩容：

```
[root@k8s-master ~]# vim nginx-Record.yaml
```

修改.spec.replicas的值：

```
spec:
  replicas: 10
[root@k8s-master ~]# kubectl apply -f nginx-Record.yaml
```

首先编辑YAML文件，将.spec.replicas的值修改为10；然后使用kubectl apply命令应用更改。

（3）通过打补丁的方式进行扩容：

```
[root@k8s-master ~]# kubectl patch deployment nginx-record -p '{"spec":{"replicas":10}}'
```

该命令用于在Kubernetes集群中直接更新名为nginx-record的Deployment的副本数（replicas）到10。kubectl patch命令允许修改资源的配置，而不需要替换整个配置文件。这对于需要快速调整资源状态（如扩展或缩减Pod数量）的场景非常有用。

2.5 StatefulSet

StatefulSet是Kubernetes中用于管理有状态服务的强大工具，它通过提供稳定的网络标识、持久化存储和有序的部署/终止等特性，使得在Kubernetes集群中部署和管理有状态服务变得更加简单和可靠。本节将介绍StatefulSet的使用方法。

2.5.1 StatefulSet 概述

StatefulSet是为了解决有状态服务的管理问题而设计的，它能够管理所有状态的服务，例如MySQL、MongoDB集群等。

Statefulset的应用场景如下。

（1）稳定的持久化存储：Pod重新调度后还是能访问到相同的持久化数据，基于PVC来实现。

（2）稳定的网络标志：Pod重新调度后其PodName和HostName不变，基于Headless Service来实现。

（3）有序部署，有序扩展：基于init containers来实现，Pod是有顺序的，在部署或扩展时要依据定义的顺序依次进行（即从0～$N-1$，在下一个Pod运行之前，所有Pod必须都是Running和Ready状态）。

（4）有序收缩，有序删除：从$N-1$到0依次删除Pod。

有状态服务和无状态服务有以下区别。

（1）有状态服务：StatefulSet是有状态的集合，用于管理有状态的服务，它所管理的Pod的名称不能随意变化。数据持久化的目录也是不一样的，每一个Pod都有自己独有的数据持久化存储目录。

（2）无状态服务：如Redis集群，由RC、Deployment、DaemonSet管理，Pod的IP、名称、启停顺序等都是随机的。个体对整体无影响，所有Pod共用一个数据卷。例如，部署的Tomcat是无状态的，Tomcat被删除后，可以启动一个新的Tomcat加入集群，与Tomcat的名称无关。

StatefulSet由以下几个部分组成。

（1）Headless Service：用来定义Pod网络标识，生成可解析的DNS记录。

（2）volumeClaimTemplates：存储卷申请模板，创建pvc，指定pvc名称大小，自动创建pvc，且pvc由存储类供应。

（3）StatefulSet：用于管理Pod。

什么是 Headless service？

Headless Service不分配Cluster IP，Headless Service可以通过解析Service的DNS返回所有Pod的地址和DNS（StatefulSet部署的Pod才有DNS），普通的Service只能通过解析Service的DNS返回Service的ClusterIP。

使用Headless Service（没有Service IP的Service）的原因：

（1）在使用Deployment时，创建的Pod名称是没有顺序的，是随机字符串。

（2）在使用StatefulSet时：

- 要求Pod名称必须是有序的，每个Pod不能被随意取代。
- Pod重建后，Pod名称保持不变。因为Pod IP是变化的，所以要使用Pod名称来识别。
- Pod名称是Pod唯一性的标识符，必须持久稳定有效。这时要用到无头服务，它可以为每个Pod提供一个唯一的名称。

2.5.2　StatefulSet 创建服务

本实例旨在验证StatefulSet在创建服务时如何确保Pod名称的有序性。通过本实例，你将观察到StatefulSet管理的Pod名称遵循预设的命名规则，这一特性对于需要状态维持的应用程序（如数据库集群）来说至关重要。

在k8s-master节点创建资源清单：

```
[root@k8s-master ~]# vim redis-sts.yaml
---
apiVersion: v1
kind: Service
metadata:
  name: redis-svc
```

```yaml
  namespace: dean
spec:
  selector:
    app: redis-sts

  ports:
  - port: 6379
    protocol: TCP
    targetPort: 6379
---
apiVersion: apps/v1
kind: StatefulSet
metadata:
  name: redis-sts
  namespace: dean

spec:
  serviceName: redis-svc
  replicas: 2
  selector:
    matchLabels:
      app: redis-sts

  template:
    metadata:
      labels:
        app: redis-sts
    spec:
      containers:
      - image: redis:5-alpine
        name: redis
        ports:
        - containerPort: 6379
```

应用资源：

```
[root@k8s-master ~]# kubectl apply -f redis-sts.yaml
service/redis-svc created
statefulset.apps/redis-sts created
```

该命令将redis-sts.yaml文件中定义的资源应用到Kubernetes集群中。使用-f选项指定要应用的文件名。根据输出信息，可以看出已经成功创建了以下资源：

- service/redis-svc created
- statefulset.apps/redis-sts created

查看Pod创建状态：

```
--- 创建第一个Pod
[root@k8s-master ~]# kubectl get pods -n dean
```

```
NAME                      READY      STATUS               RESTARTS       AGE
redis-sts-0               0/1        ContainerCreating    0              16s
```

--- 第一个Pod启动完成才创建下一个

```
[root@k8s-master ~]# kubectl get pods -n dean
NAME                      READY      STATUS               RESTARTS       AGE
redis-sts-0               1/1        Running              0              32s
redis-sts-1               0/1        ContainerCreating    0              7s
```

--- Pod全部运行成功

```
[root@k8s-master ~]# kubectl get pods -n dean
NAME           READY    STATUS    RESTARTS    AGE
redis-sts-0    1/1      Running   0           55s
redis-sts-1    1/1      Running   0           30s
```

通过实践，你将深刻理解StatefulSet如何满足这些复杂应用的部署需求，确保每个Pod都能以预期的顺序和命名规范被创建和管理。

2.5.3 Ping 域名实验

本实例用于测试Kubernetes中StatefulSet的资源管理特性，通过Kubectl工具直接访问名为redis-sts-0的Pod内部，并尝试对该StatefulSet中的另一个Pod，即redis-sts-1，通过名为redis-svc的服务地址执行ping操作。实验结果表明，两个Pod之间通过服务名能够成功通信，验证了StatefulSet中Pod间的网络互通性以及服务的有效性。

接下来，实例进一步模拟了Pod故障恢复的场景，通过删除名为redis-sts-1的Pod，观察StatefulSet自动重建Pod的过程。在重建完成后，再次执行对redis-sts-1的ping操作，发现其IP地址已更新，但仍能通过服务名成功访问，这展示了StatefulSet在Pod恢复时的自动IP分配与更新机制，以及服务发现功能的稳定性。通过这些步骤，实验全面验证了StatefulSet在Kubernetes中对于状态维护型应用的高可用性和弹性伸缩能力。

查看Service信息：

```
[root@k8s-master ~]# kubectl get svc -n dean
NAME         TYPE        CLUSTER-IP       EXTERNAL-IP    PORT(S)      AGE
redis-svc    ClusterIP   10.104.5.170     <none>         6379/TCP     172m
```

进入redis-sts-0这个Pod进行ping redis-sts-1：

```
[root@k8s-master ~]# kubectl exec -it -n dean redis-sts-0 -- sh
/data # ping redis-sts-1.redis-svc
PING redis-sts-1.redis-svc (10.244.36.80): 56 data bytes
64 bytes from 10.244.36.80: seq=0 ttl=62 time=0.165 ms
64 bytes from 10.244.36.80: seq=1 ttl=62 time=0.286 ms
^C
--- redis-sts-1.redis-svc ping statistics ---
6 packets transmitted, 6 packets received, 0% packet loss
```

```
round-trip min/avg/max = 0.165/0.246/0.288 ms
/data #
```

该命令使用Kubectl工具进入名为redis-sts-0的Pod内，对名为redis-sts-1的Pod且名为redis-svc的Service进行ping。根据输出信息，可以看出这两个Pod是互通的。

删除redis-sts-1这个Pod：

```
[root@k8s-master ~]# kubectl delete pods -n dean redis-sts-1
pod "redis-sts-1" deleted
```

该命令使用Kubectl工具删除Kubernetes集群中dean命名空间下名为redis-sts-1的Pod。根据输出信息，可以看出已经删除了以下资源：

```
pod "redis-sts-1" deleted
```

查看Pod资源信息：

```
[root@k8s-master ~]# kubectl get pods -o wide -n dean
NAME          READY   STATUS    RESTARTS   AGE     IP               NODE        NOMINATED NODE   READINESS GATES
redis-sts-0   1/1     Running   0          4h57m   10.244.107.207   k8s-node3   <none>           <none>
redis-sts-1   1/1     Running   0          4s      10.244.36.81     k8s-node1   <none>           <none>
```

再次进入redis-sts-0这个Pod进行ping redis-sts-1：

```
[root@k8s-master ~]# kubectl exec -it -n dean redis-sts-0 -- sh
/data # ping redis-sts-1.redis-svc
PING redis-sts-1.redis-svc (10.244.36.81): 56 data bytes
64 bytes from 10.244.36.81: seq=0 ttl=62 time=0.271 ms
64 bytes from 10.244.36.81: seq=1 ttl=62 time=0.301 ms
64 bytes from 10.244.36.81: seq=2 ttl=62 time=0.332 ms
^C
--- redis-sts-1.redis-svc ping statistics ---
3 packets transmitted, 3 packets received, 0% packet loss
round-trip min/avg/max = 0.271/0.301/0.332 ms
```

根据输出信息可以得出，当Pod重建后，IP地址会发生变更，Service会自动解析到新的Pod IP上。

2.5.4 解析域名实验

本实例通过运用dig工具，针对Kubernetes集群中部署的CoreDNS服务的IP地址发起DNS查询请求，目标是对特定Service进行域名解析。实验结果显示，CoreDNS成功地将Service名称解析为后端Pod的IP地址，这一过程不仅验证了CoreDNS作为Kubernetes集群内部DNS解决方案的有效性，还展示了Service与Pod之间通过DNS进行服务发现与通信的机制。此实验深入探索了Kubernetes网络模型中的服务发现机制，确保了集群内部组件能够高效、准确地相互定位与通信。

StatefulSet 使用 Headless 服务来控制 Pod 的域名，这个域名的 FQDN 为 $(service name).$(namespace).svc.cluster.local，其中，cluster.local指的是集群的域名。

查看CoreDNS的IP地址：

```
[root@k8s-master ~]# kubectl get pods -o wide -n kube-system|grep core
coredns-6d8c4cb4d-frw8v               1/1     Running   1          10d
10.244.107.194    k8s-node3    <none>           <none>
coredns-6d8c4cb4d-lnfsp               1/1     Running   1 (10d ago) 10d
10.244.36.67      k8s-node1    <none>           <none>
```

该命令是在kube-system命名空间中查找所有Pod，并以更宽的格式显示它们的详细信息，然后过滤出名称中包含core字符串的Pod。从输出信息可以看出，名为coredns的Pod被过滤展示出来了。

使用K8s集群内部DNS解析Pod域名：

如果找不到dig命令，则执行yum -y install bind-utils进行安装。

```
[root@k8s-master ~]# dig -t A redis-svc.dean.svc.cluster.local. @10.244.36.67

; <<>> DiG 9.11.4-P2-RedHat-9.11.4-26.P2.el7_9.15 <<>> -t A
redis-svc.dean.svc.cluster.local. @10.244.36.67
;; global options: +cmd
;; Got answer:
;; WARNING: .local is reserved for Multicast DNS
;; You are currently testing what happens when an mDNS query is leaked to DNS
;; ->>HEADER<<- opcode: QUERY, status: NOERROR, id: 4216
;; flags: qr aa rd; QUERY: 1, ANSWER: 1, AUTHORITY: 0, ADDITIONAL: 1
;; WARNING: recursion requested but not available

;; OPT PSEUDOSECTION:
; EDNS: version: 0, flags:; udp: 4096
;; QUESTION SECTION:
;redis-svc.dean.svc.cluster.local.   IN  A

;; ANSWER SECTION:
redis-svc.dean.svc.cluster.local. 30 IN A     10.104.5.170

;; Query time: 0 msec
;; SERVER: 10.244.36.67#53(10.244.36.67)
;; WHEN: 二 5月 14 16:44:59 CST 2024
;; MSG SIZE  rcvd: 109
```

该命令使用dig工具，向IP地址为10.244.36.67的DNS服务器查询redis-svc.dean.svc.cluster.local.这个域名的A记录（即IPv4地址）。这通常用于在Kubernetes集群内部查找服务的IP地址，以便进行服务间通信。

2.5.5 创建 StatefulSet 服务自动申请 PV 实验

本实例精心设计了一个关于StatefulSet服务的部署过程，其中涵盖自动申请PersistentVolume（PV）以持久化存储数据的功能。随后，利用Nslookup工具对StatefulSet中的Pod进行了DNS解析测试，验证了集群内部DNS服务的正确性与高效性，确保了Pod能够通过其服务名在集群网络中被准确识别与访问。此实验不仅展示了StatefulSet在管理有状态应用方面的强大能力，还强调了Kubernetes集群中DNS服务对于服务发现与通信的关键作用。

01 在k8s-master节点创建资源清单：

```yaml
[root@k8s-master ~]# vim nginx-sts.yaml
---
apiVersion: v1
kind: Service
metadata:
  name: nginx
  namespace: dean
  labels:
    app: nginx
spec:
  ports:
  - port: 80
    name: web
  clusterIP: None
  selector:
    app: nginx
---
apiVersion: apps/v1
kind: StatefulSet
metadata:
  name: nginx
  namespace: dean
spec:
  selector:
    matchLabels:
      app: nginx
  serviceName: "nginx"
  replicas: 2
  template:
    metadata:
      labels:
        app: nginx
    spec:
      containers:
      - name: nginx
        image: nginx:1.19
        ports:
        - containerPort: 80
          name: web
        volumeMounts:
        - name: www
          mountPath: /usr/share/nginx/html
```

```
      # volumeClaimTemplates 通常与StatefulSet一起使用，用于为StatefulSet中的每个Pod创建独立
的持久化存储卷
      volumeClaimTemplates:
      - metadata:
          # 存储卷声明的名称为www。这意味着StatefulSet中的每个Pod都将获得一个名为www的持久化存储卷
          name: www
          # 指定了存储类为nfs。存储类（StorageClass）是Kubernetes中用于描述存储"类"的API对象，
它允许管理员描述存储的"类型"或"性能级别"，如SSD、HDD或NFS等。这里指定NFS意味着存储后端将使用NFS（网
络文件系统）来提供存储
          annotations:
              volume.beta.kubernetes.io/storage-class: "nfs"
        spec:
          # 指定存储卷的访问模式为ReadWriteOnce，意味着该卷可以被单个节点以读写模式挂载。这是StatefulSet中
常见的访问模式，因为StatefulSet通常用于需要稳定且持久存储的应用，如数据库或需要文件存储的应用
          accessModes: ["ReadWriteOnce"]
          resources:
            requests:
              # 表示请求的存储大小为1GiB（吉字节）。这是Pod启动时向存储提供者请求的最小存储空间量
              storage: 1Gi
```

02 应用资源：

```
[root@k8s-master ~]# kubectl apply -f nginx-sts.yaml
service/nginx created
statefulset.apps/nginx created
```

该命令使用Kubectl工具将nginx-sts.yaml文件中定义的资源应用到Kubernetes集群中。使用-f选项指定要应用的文件名。从输出信息可以看出，已经成功创建了以下资源：

- service/nginx created
- statefulset.apps/nginx created

03 查看Pod资源状态：

```
[root@k8s-master ~]# kubectl get pods -n dean
NAME       READY   STATUS    RESTARTS   AGE
nginx-0    1/1     Running   0          17m
nginx-1    1/1     Running   0          17m
```

该命令使用Kubectl工具查看在Kubernetes集群中dean命名空间下的Pod资源信息。

04 查看Headless Service：

```
[root@k8s-master ~]# kubectl get svc -n dean
NAME    TYPE        CLUSTER-IP   EXTERNAL-IP   PORT(S)   AGE
nginx   ClusterIP   None         <none>        80/TCP    18m
```

该命令使用Kubectl工具查看在Kubernetes集群中dean命名空间下的Service资源信息。

05 查看PV：

```
[root@k8s-master ~]# kubectl get pv
NAME                                       CAPACITY   ACCESS MODES   RECLAIM POLICY   STATUS   CLAIM           STORAGECLASS   REASON   AGE
pvc-066aecdd-13cd-4cb3-97de-3494fde06395   1Gi        RWO            Delete           Bound    dean/nginx-1    nfs                     19m
pvc-f178e28f-42d4-45f5-a53e-e71ee954a840   1Gi        RWO            Delete           Bound    dean/nginx-0    nfs                     19m
```

该命令使用Kubectl工具查看在Kubernetes集群中的PV资源信息。

06 查看PVC：

```
[root@k8s-master ~]# kubectl get pvc -n dean
NAME      STATUS   VOLUME                                     CAPACITY   ACCESS MODES   STORAGECLASS   AGE
nginx-0   Bound    pvc-f178e28f-42d4-45f5-a53e-e71ee954a840   1Gi        RWO            nfs            20m
nginx-1   Bound    pvc-066aecdd-13cd-4cb3-97de-3494fde06395   1Gi        RWO            nfs            20m
```

该命令使用Kubectl工具查看在Kubernetes集群中dean命名空间下的PVC资源信息。

07 查看Pod主机名：

```
[root@k8s-master ~]# for i in 0 1; do kubectl exec -n dean nginx-$i -- sh -c 'hostname'; done
# 输出信息
nginx-0
nginx-1
```

该命令在Kubernetes集群的dean命名空间中对名为nginx-0和nginx-1的Pod执行hostname命令。

08 使用kubectl run运行一个提供nslookup命令的容器，该命令来自dnsutils包，通过对Pod主机名执行nslookup，可以检查它们在集群内部的DNS地址：

```
[root@k8s-master ~]# kubectl exec -n dean -it nginx-1 -- /bin/bash
root@nginx-1:/# apt-get update
root@nginx-1:/# apt-get install dnsutils -y
root@nginx-1:/# nslookup nginx-0.nginx.dean.svc.cluster.local
Server:         10.96.0.10
Address:        10.96.0.10#53

Name:   nginx-0.nginx.dean.svc.cluster.local
Address: 10.244.169.155

root@web-1:/# nslookup nginx-0.nginx
Server:         10.96.0.10
Address:        10.96.0.10#53

Name:   nginx-0.nginx.dean.svc.cluster.local
Address: 10.244.169.155

# statefulset创建的Pod也是有DNS记录的
Address: 10.244.169.155  # 解析的是Pod的IP地址
```

2.6 DaemonSet

本节将介绍DaemonSet控制器的使用方法。

2.6.1 DaemonSet 概述

DaemonSet控制器在Kubernetes中具有重要作用,其主要功能包括确保每个节点上运行指定的Pod副本、自动管理Pod的生命周期、提供滚动更新和自动恢复等。

DaemonSet通过以下几个关键功能来实现其作用。

- 确保每个节点上运行指定的Pod副本:无论集群规模如何,DaemonSet都会在每个节点上部署指定数量的Pod副本。当有新节点加入集群时,DaemonSet也会自动在新节点上创建相应的Pod,保证服务覆盖整个集群。
- 自动管理Pod的生命周期:DaemonSet控制器不仅在节点加入时创建Pod,还会监视这些Pod的运行状况。如果某个Pod因节点故障或其他原因停止运行,DaemonSet会尝试重新创建该Pod,以保持服务的持续性。
- 提供滚动更新和自动恢复:当DaemonSet的配置或模板更新时,它支持滚动更新操作,逐步替换旧Pod,以减少对正在运行的服务的影响。同时,如果出现更新失败的情况,DaemonSet还支持自动回滚到之前的状态。
- 提供高度灵活性和可定制性:DaemonSet支持多种调度策略,如NodeSelector和NodeAffinity,允许用户根据节点的标签或属性来约束Pod运行的节点。这提供了极大的灵活性,使得DaemonSet可以适应各种复杂的集群环境和需求。
- 集成其他Kubernetes资源:DaemonSet紧密集成了Kubernetes的其他核心资源,如Service、Volumes和Network Policies,这使得开发者可以统一配置和管理集群内的各类资源。
- 优化资源利用:DaemonSet允许设定资源请求和限制,以确保Pod在节点上的资源使用符合预期,避免资源争抢和浪费。

利用上述特性,我们可以使用DaemonSet部署Fluentd作为日志收集器,或者部署Prometheus Node Exporter作为监控代理等。典型用法如下:

(1)运行集群存储Daemon,例如在每个Node上运行glusterd、ceph。

(2)在每个Node上运行日志收集Daemon,例如Fluentd、Logstash ELK。

(3)在每个Node上运行监控Daemon,例如Prometheus Node Exporter、Collectd、Datadog代理、New Relic代理或Ganglia gmond。

DaemonSet在Kubernetes中的使用不仅提高了集群服务的可靠性和稳定性,还简化了集群的管理工作,合理使用DaemonSet可以显著提升集群的整体性能和运维效率。

2.6.2 DaemonSet 实例

本实例旨在展示DaemonSet如何自动在每个Kubernetes集群节点上部署并管理Pod实例。在实验过程中，当手动删除某个节点上的DaemonSet管理的Pod时，DaemonSet的控制器会自动在该节点上重新创建一个新的Pod，以确保每个节点上始终运行着指定的Pod副本。这一特性体现了DaemonSet在集群节点级服务部署与维护中的独特价值，确保了集群中每个节点都能提供一致的服务能力。

01 在k8s-master节点创建资源清单：

```
[root@k8s-master ~]# vim daemonset.yaml
apiVersion: apps/v1
kind: DaemonSet
metadata:
  name: daemonset-example
  namespace: dean
  labels:
    app: daemonset
spec:
  selector:
    matchLabels:
      name: daemonset-example
  template:
    metadata:
      labels:
        name: daemonset-example
    spec:
      containers:
      - name: daemonset-example
        image: nginx:1.21
        imagePullPolicy: IfNotPresent
        ports:
          - containerPort: 80
```

02 应用资源：

```
[root@k8s-master ~]# kubectl apply -f daemonset.yaml
daemonset.apps/daemonset-example created
```

该命令使用Kubectl工具将daemonset.yaml文件中定义的资源应用到Kubernetes集群中。使用-f选项指定要应用的文件名。从输出信息可以看出，已经成功创建了以下资源：

```
daemonset.apps/daemonset-example created
```

03 查看Pod资源的详细信息，可看到每个Node节点都有一个Pod：

```
[root@k8s-master ~]# kubectl get pods -n dean -o wide
     NAME                          READY   STATUS     RESTARTS   AGE     IP
NODE         NOMINATED NODE    READINESS GATES
```

```
    daemonset-example-796pt              1/1    Running    0    53s
10.244.107.242    k8s-node3    <none>    <none>
    daemonset-example-ptk68              1/1    Running    0    53s
10.244.169.161    k8s-node2    <none>    <none>
    daemonset-example-w8glk              1/1    Running    0    53s
10.244.36.115     k8s-node1    <none>    <none>
```

04 更新镜像：

```
[root@k8s-master ~]# kubectl set image -n dean ds daemonset-example
daemonset-example=nginx:1.22
daemonset.apps/daemonset-example image updated
```

05 查看Pod的Nginx版本，发现已经被更新：

```
[root@k8s-master ~]# kubectl exec -it -n dean daemonset-example-25qgt -- nginx -v
nginx version: nginx/1.22.0
```

06 使用DaemonSet控制器删除任意一个Pod，也会在原节点创建新的Pod。

2.7 CronJob

CronJob控制器允许用户以Cron表达式的方式设置特定的时间点或时间间隔来执行任务，即基于时间的调度在指定时间自动运行Job。本节将介绍CronJob的使用方法。

2.7.1 CronJob 概述

CronJob是Kubernetes中用来管理定时任务的资源对象，它通过创建Job对象来实现其功能，每个Job再根据需要创建一个或多个Pod。其作用和功能如下。

- 周期性地执行任务：CronJob的核心功能是周期性地执行任务。用户可以定义具体的Cron表达式来指定任务的执行频率，例如每天、每周或每月等。这大大简化了对周期性任务的管理，使得运维人员不再需要手动启动任务，系统会自动按照预定计划进行操作。
- 定时与灵活控制：CronJob支持精确到秒的定时调度，并允许复杂的时间表达式，如"0 0 * * *"表示每小时执行一次。这种灵活性使得CronJob适用于各种复杂的调度需求，比如在业务低峰期执行数据备份，或者在每天的特定时间进行系统维护检查。
- 失败重试与自我修复：当Job因为故障未能成功完成时，CronJob可以配置重试策略，重新创建Job来尝试完成任务。这种自我修复机制确保了任务的高可靠性，即使在某些极端情况下，也能保证任务最终得以执行。
- 资源清理与历史记录：CronJob可以配置保留成功或失败Job的历史记录，方便追踪和分析。同时，可以通过设置successfulJobsHistoryLimit和failedJobsHistoryLimit参数来控制保留的历史数量，防止过多的旧任务占用系统资源。

- 暂停与恢复：在某些情况下，可能需要暂时停止CronJob的执行，CronJob提供了suspend功能，可以通过修改spec.suspend为true来暂停任务的执行，当需要恢复时，再将其设置为false。这样，在不删除CronJob的情况下，可以临时停止任务调度。
- 并发控制与时区设置：CronJob支持设置并发策略，如Allow、Forbid和Replace，以控制新任务与正在运行中的任务的交互方式。同时，CronJob还支持时区设置，这对于跨时区部署的集群非常重要，能够确保任务按时执行。

CronJob通过提供高度灵活和自动化的方式来执行周期性任务，大大提高了系统的可维护性和可靠性，合理使用CronJob可以显著提升集群的整体性能和运维效率。

请注意，CronJob是在Kubernetes 1.8版本中引入的，因此，只有在1.8及以上版本的Kubernetes集群中才可以使用CronJob。

CronJob的调度执行是基于controller-manager的系统时间。

CronJob控制器以Job控制器资源为其管理对象，并利用Job来管理和控制Pod资源，通常缩写为CJ。

Job控制器定义的作业任务在其控制器资源创建之后便会立即执行，但CronJob能够以类似于Linux操作系统的周期性任务作业计划的方式，控制其运行时间点及重复运行的方式。

也就是说，CronJob可以在特定的时间点（反复的）运行Job任务。

CronJob控制器调用图如图2-1所示。

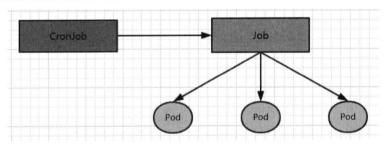

图 2-1　CronJob 控制器调用图

2.7.2　CronJob 资源清单详解

资源清单模板：

```
apiVersion: batch/v1         # 版本号
kind: CronJob                # 类型
metadata:                    # 元数据
  name: hello                # Cronjob名称
  namespace: dean            # 命名空间
  labels:                    # 标签
    controller: cronjob
```

```yaml
spec:                              # 详情描述
  concurrencyPolicy: Allow    # 并发执行策略,用于定义前一次作业运行尚未完成时是否以及如何运行后一次作业
  #Allow(默认):允许并发运行Job
  #Forbid:禁止并发运行,如果前一个还没有完成,则直接跳过下一个
  #Replace:取消当前正在运行的Job,用一个新的来替换
  ###############################################################
  successfulJobsHistoryLimit: 3      # 要保留的成功完成的作业数(默认为3)
  failedJobsHistoryLimit: 1          # 为失败的任务执行保留的历史记录数,默认为1
  schedule: '*/1 * * * *'            # 作业时间表,作业将每分钟运行一次(分时日月周)
  startingDeadlineSeconds: 3         # Pod必须在规定时间后的3秒内开始执行,若超过该时间未执行,则任务将不运行,且标记失败
  suspend: false            # true表示挂起不运行,false表示挂起并运行
  jobTemplate:              # Job控制器模板,用于为CronJob控制器生成Job对象;下面其实就是Job的定义
    metadata:
      labels:
        controller: cronjob
    spec:
      completions: 1           # 标志Job结束需要成功运行的Pod个数,默认为1
      parallelism: 1           # 标志并运行的Pod的个数,默认为1
      activeDeadlineSeconds: 1 # 标志失败Pod的重试最大时间,超过这个时间不会继续重试
      template:
        metadata:
          labels:
            controller: cronjob
        spec:
          dnsPolicy: ClusterFirst
          restartPolicy: OnFailure    # 重启策略
          schedulerName: default-scheduler
          securityContext: {}
          terminationGracePeriodSeconds: 30
          containers:
          - args:
            - /bin/sh
            - -c
            - date; echo Hello I am Dean
            image: busybox
            imagePullPolicy: IfNotPresent
            name: hello
            terminationMessagePath: /dev/termination-log
            terminationMessagePolicy: File
```

使用kubectl explain cronjob来查看当前集群版本的资源对象。

2.7.3 CronJob 实验

本实例用于CronJob的创建与功能验证,旨在通过配置CronJob来自动化地执行特定命令,设定任务每隔一分钟触发一次,以此展示Kubernetes中CronJob资源在定时任务调度方面的强大能力。

01 在k8s-master节点创建资源清单：

```
[root@k8s-master ~]# vim cronjob.yaml
apiVersion: batch/v1
kind: CronJob
metadata:
  name: hello
  namespace: dean
spec:
  schedule: "*/1 * * * *"
  jobTemplate:
    spec:
      template:
        spec:
          restartPolicy: OnFailure
          containers:
            - name: hello
              imagePullPolicy: IfNotPresent
              image: busybox
              args:
                - /bin/sh
                - -c
                - date; echo Hello I am Dean
```

02 应用资源：

```
[root@k8s-master ~]# kubectl apply -f cronjob.yaml
cronjob.batch/hello created
```

03 使用Kubectl工具将cronjob.yaml文件中定义的资源应用到Kubernetes集群中。使用-f选项指定要应用的文件名。从输出信息可以看出，已经成功创建了以下资源：

```
cronjob.batch/hello created
```

04 查看Job，发现已经有两个，每一分钟创建一个：

```
[root@k8s-master ~]# kubectl get job -n dean
NAME                COMPLETIONS   DURATION   AGE
hello-1606808040    1/1           2s         81s
hello-1606808100    1/1           2s         21s
```

05 查看CronJob：

```
[root@k8s-master ~]# kubectl get cronjob -n dean
NAME    SCHEDULE      SUSPEND   ACTIVE   LAST SCHEDULE   AGE
hello   */1 * * * *   False     0        35s             101s
```

参数解析：

- NAME：显示CronJob的名称，这里是hello。

- SCHEDULE：显示CronJob的调度计划。这里的*/1 * * * *表示每分钟的每一秒都会触发一次（但实际上在Kubernetes CronJob中，通常不会精确到秒，这里可能是个误解或者特殊用法。标准的Kubernetes CronJob调度格式是基于分钟的，如*/1 * * * *意味着每分钟执行一次）。
- SUSPEND：显示CronJob是否被挂起。False表示CronJob当前没有被挂起，正常调度执行。
- ACTIVE：显示当前活跃的Job数量。0表示没有活跃的Job。
- LAST SCHEDULE：显示上一次调度执行的时间。这里显示35s前执行了上一次调度。
- AGE：显示CronJob创建以来的时间。这里显示101s，意味着这个CronJob已经创建了101秒。

06 查看Pod，状态为Completed，两个Pod的时间相差一分钟：

```
[root@k8s-master ~]# kubectl get pods -n dean
NAME                       READY   STATUS      RESTARTS   AGE
hello-1606808040-njjsj     0/1     Completed   0          94s
hello-1606808100-qvk9l     0/1     Completed   0          34s
```

07 查看Pod日志：

```
--- 两个Pod输出内容的时间也是相差1分钟
[root@k8s-master ~]# kubectl logs hello-1606808040-njjsj -n dean
Tue May 21 05:21:00 UTC 2024
Hello I am Dean

[root@k8s-master ~]# kubectl logs hello-1606808100-qvk9l -n dean
Tue May 21 05:22:00 UTC 2024
Hello I am Dean
```

08 删除CronJob时不会自动删除Job，可以用kubectl delete job来删除源对象。

2.7.4 CronJob 实战备份 MySQL 数据库

实验思路：启用一个CronJob，使用的镜像是mysql:5.7（注意密码必须设置，所以用到了MYSQL_ROOT_PASSWORD），将一个持久卷挂载至Pod内，使用mysqldump工具备份数据库至持久卷目录，这样备份的数据库文件将会存储到NFS服务器中。

01 在k8s-master节点创建PVC资源清单：

```
[root@k8s-master ~]# vim mysqldump-cronjob-pvc.yaml
apiVersion: v1
kind: PersistentVolumeClaim
metadata:
  name: mysqldump-cronjob-pvc
  namespace: dean
spec:
  storageClassName: "nfs-storage"
  accessModes:
    - ReadWriteMany
  resources:
    requests:
```

```
          storage: 1Gi
```

02 在k8s-master节点创建CronJob资源清单:

```
[root@k8s-master ~]# vim mysqldump-cronjob.yaml
apiVersion: batch/v1
kind: CronJob
metadata:
  name: mysqldump-cronjob
  namespace: dean
spec:
  schedule: "*/1 * * * *"
  jobTemplate:
    spec:
      template:
        spec:
          containers:
          - name: mysqldump
            image: mysql:5.7
            env:
            # MySQL密码
            - name: MYSQL_ROOT_PASSWORD
              value: dean
            # 执行命令,使用mysqldump工具连接数据库,对名为dean的库进行备份,备份至/mnt目录
            command: ["/bin/sh","-c","mysqldump --host=192.168.1.20 -P3306 -uroot -pdeanmysql --databases dean >/mnt/dean`date +%Y%m%d%H%M`.sql"]
            # 卷挂载
            volumeMounts:
              - name: cronjob-pvc
                # 挂载至Pod的目录
                mountPath: "/mnt"
          restartPolicy: Never
          # 卷信息
          volumes:
            - name: cronjob-pvc
              # 使用PVC
              persistentVolumeClaim:
                # PVC名字
                claimName: mysqldump-cronjob-pvc
```

03 应用资源:

```
[root@k8s-master ~]# kubectl apply -f mysqldump-cronjob-pvc.yaml
persistentvolumeclaim/mysqldump-cronjob-pvc created
[root@k8s-master ~]# kubectl apply -f mysqldump-cronjob.yaml
cronjob.batch/mysqldump-cronjob created
```

04 该命令使用Kubectl工具将mysqldump-cronjob-pvc.yaml和mysqldump-cronjob.yaml文件中定义的资源应用到Kubernetes集群中。使用-f选项指定要应用的文件名。从输出信息可以看出,已经成功创建了以下资源:

```
persistentvolumeclaim/mysqldump-cronjob-pvc created
cronjob.batch/mysqldump-cronjob created
```

05 查看Pod状态：

--- 查看cj状态
```
[root@k8s-master ~]# kubectl get cj -n dean
NAME                SCHEDULE      SUSPEND   ACTIVE   LAST SCHEDULE   AGE
mysqldump-cronjob   */1 * * * *   False     0        35s             91s
```

--- 查看Pod状态
```
[root@k8s-master ~]# kubectl get pods -n dean
NAME                                  READY   STATUS      RESTARTS   AGE
mysql-dump-l2q6t                      0/1     Completed   0          87m
mysqldump-cronjob-1644911520-v5h8c    0/1     Completed   0          30s
```

--- 查看cj状态，ACTIVE为0，LAST时间为69秒，60秒执行一次定时任务
```
[root@k8s-master ~]# kubectl get cj -n dean
NAME                SCHEDULE      SUSPEND   ACTIVE   LAST SCHEDULE   AGE
mysqldump-cronjob   */1 * * * *   False     0        69s             2m5s
```
--- 再次查看cj状态，可以看到ACTIVE为1，LAST时间已经重新开始计算
```
[root@k8s-master ~]# kubectl get cj -n dean
NAME                SCHEDULE      SUSPEND   ACTIVE   LAST SCHEDULE   AGE
mysqldump-cronjob   */1 * * * *   False     1        10s             2m6s
```

--- 查看Pod状态，发现定时任务Pod已经开始创建
```
[root@k8s-master ~]# kubectl get pods -n dean
NAME                                  READY   STATUS              RESTARTS   AGE
mysql-dump-l2q6t                      0/1     Completed           0          88m
mysqldump-cronjob-1644911520-v5h8c    0/1     Completed           0          64s
mysqldump-cronjob-1644911580-8svsk    0/1     ContainerCreating   0          3s
```

--- 查看Pod状态，定时任务Pod已经正在运行，此时正在备份数据库
```
[root@k8s-master ~]# kubectl get pods -n dean
NAME                                  READY   STATUS      RESTARTS   AGE
mysql-dump-l2q6t                      0/1     Completed   0          88m
mysqldump-cronjob-1644911520-v5h8c    0/1     Completed   0          68s
mysqldump-cronjob-1644911580-8svsk    1/1     Running     0          7s
```

--- 查看Pod状态，定时任务Pod已经完成，数据库备份完成
```
[root@k8s-master ~]# kubectl get pods -n dean
NAME                                  READY   STATUS      RESTARTS   AGE
mysql-dump-l2q6t                      0/1     Completed   0          88m
mysqldump-cronjob-1644911520-v5h8c    0/1     Completed   0          68s
mysqldump-cronjob-1644911580-8svsk    0/1     Completed   0          7s
```

06 进入NFS的存储目录，查看备份的SQL文件，可以看到时间为一分钟：

```
[root@k8s-master
dean-mysqldump-cronjob-pvc-pvc-8a8cf08b-74f6-4bcc-af1f-4abfddd38669]# ls -lht
总用量 17M
```

```
-rw-r--r--. 1 root root 4.1M 5月  21 15:55 dean202405211555.sql
-rw-r--r--. 1 root root 4.1M 5月  21 15:54 dean202405211554.sql
-rw-r--r--. 1 root root 4.1M 5月  21 15:53 dean202405211553.sql
-rw-r--r--. 1 root root 4.1M 5月  21 15:52 dean202405211552.sql
```

此实验用到的通过存储类申请PV，将在后面章节进行介绍。

2.8 Job

Job控制器用于执行一次性任务或批处理作业，可以用来监视Job对象的状态，例如根据需要启停Pod等。本节将介绍Job控制器的使用方法。

2.8.1 Job概述

Kubernetes Job是一个控制器，它可以创建和管理一次性任务。当Job被创建时，它将自动创建一个或多个Pod来运行任务，直到任务成功完成为止。

通过修改Job Spec中的.spec.suspend字段可以实现暂停Job，用户可以在需要时再次将其设置为false，恢复Job的执行。

与其他控制器不同，Job控制器负责确保Pod成功地完成任务后删除它们。这意味着如果任务失败或调度失败（例如，由于资源不足），Kubernetes将自动重新启动失败的Pod，直到任务成功完成为止。

Job通常用于数据处理、备份和恢复操作等场景。例如，从数据库中导出数据到CSV文件、打包和压缩日志文件、备份数据库和配置文件以及执行机器学习模型的特征向量计算等。

2.8.2 Job实验

本实验旨在通过Kubernetes平台创建并运行一个Job任务，该任务负责执行计算圆周率（π）小数点后2000位的复杂计算过程。这一实验不仅展示了Kubernetes在一次性或批处理任务调度与执行方面的灵活性，还体现了其资源管理与调度机制的强大功能。

01 在k8s-master节点创建资源清单：

```
[root@k8s-master ~]# vim job.yaml
apiVersion: batch/v1
kind: Job
metadata:
  name: pi
  namespace: dean
spec:
  template:
    metadata:
      name: pi
```

```yaml
spec:
  containers:
    - name: pi
      image: perl
      imagePullPolicy: IfNotPresent
      # 计算π后边的2000位
      command: ["perl","-Mbignum=bpi","-wle","print bpi(2000)"]
  restartPolicy: Never
```

02 应用资源：

```
[root@k8s-master ~]# kubectl apply -f job.yaml
job.batch/pi created
```

03 该命令使用Kubectl工具将job.yaml文件中定义的资源应用到Kubernetes集群中。使用-f选项指定要应用的文件名。从输出信息可以看出，已经成功创建了以下资源：

```
job.batch/pi created
```

04 查看Pod资源状态：

```
--- 正在创建Pod
[root@k8s-master ~]# kubectl get pods -n dean
NAME          READY   STATUS              RESTARTS   AGE
pi-r5g58      0/1     ContainerCreating   0          4s
--- 查看job，COMPLETIONS为0
[root@k8s-master ~]# kubectl get job -n dean
NAME   COMPLETIONS   DURATION   AGE
pi     0/1           25s        25s

--- Pod中的任务执行完毕，状态为Completed
[root@k8s-master ~]# kubectl get pods -n dean
NAME          READY   STATUS       RESTARTS   AGE
pi-r5g58      0/1     Completed    0          11m
--- 再次查看job，COMPLETIONS为1
[root@k8s-master ~]# kubectl get job -n dean
NAME   COMPLETIONS   DURATION   AGE
pi     1/1           86s        13m
```

05 查看Pod日志：

```
[root@k8s-master ~]# kubectl logs pi-r5g58 -n dean
3.1415926535897932384626433832795028841971693993751058209749445923078164062862089
9862803482534211706798214808651328230664709384460955058223172535940812848111
...此处省略若干...
17450284102896084128488626945604241965285022210661186306744278622039194945047123
7137869609563643719172874677646575739624138908658326459958133904780275901
```

2.9　本章小结

本章通过介绍Kubernetes中的Pod控制器，展示了Kubernetes在自动化和高效管理大规模容器化应用方面的强大能力。这些控制器不仅简化了Pod的管理流程，还提供了丰富的功能来满足不同应用场景的需求。无论是需要频繁更新的应用、需要保持状态一致性的有状态应用，还是需要在每个节点上运行的服务，Kubernetes都提供了相应的控制器来支持。掌握这些控制器的使用，对于构建和管理高效、可靠的Kubernetes集群至关重要。

通过本章的学习和实践，读者应该可以对Kubernetes的Pod控制器有一个深入的了解，为后续章节的学习打下基础。

第 3 章

Label、容器钩子、探针

本章主要介绍关于Kubernetes中的三个关键组件——Label、容器钩子及探针。首先深入探讨Label的概念,揭示它是如何作为连接Kubernetes资源与操作策略的纽带,助力用户高效组织、筛选和管理集群中的各类对象的。随后,将转向容器钩子的解析,展示这些在容器生命周期关键阶段自动触发的操作如何为开发者提供额外的控制力,优化应用的部署、运行和清理流程。最后,详细阐述探针的类型与工作原理,包括存活探针、就绪探针和启动探针,帮助读者理解这些机制如何确保容器健康,提高服务可用性和可靠性。

3.1 Label标签

Label是附加到Kubernetes对象上的键－值对,它不仅可以帮助用户识别和组织资源,还支持复杂的资源选择和操作。本节将介绍Label的概念和分类,并通过实验来掌握其应用。

3.1.1 Label 概述

在Kubernetes(K8s)中,Label(标签)作为一种强大的工具,允许用户将自定义的元数据附加到各种资源对象上,如Pods、Services、Deployments等。通过Label,用户可以方便地组织和筛选这些对象进行批量的操作和管理。简而言之,Label由用户定义的键(key)和值(value)组成,它们为Kubernetes对象提供了关键的识别属性。

Label有多种类型,分别说明如下。

- 版本标签(release):例如stable(稳定版)、canary(金丝雀版本)、beta(测试版)。
- 环境类(environment):例如ev(开发)、qa(测试)、production(生产)、op(运维)。
- 应用类(application):例如ui(设计)、as(应用软件)、pc(计算机端)、sc(网络)。
- 架构层(tier):例如frontend(前端)、backend(后端)、cache(缓存)。

- 分区标签（partition）：例如customerA、customerB。
- 品控级别（track）：例如daily（每天）、weekly（每周）。

我们可以根据实际需要来选择使用不同的Label。

3.1.2 Label 实验

本实验的目标是通过实际操作演示如何在创建Deployment控制器资源时，为该Deployment管理的Pod设置特定的Label。本实验将进一步指导如何查询带有特定Label的Pod，并最终演示如何移除这些Label。

实验步骤如下：

01 在k8s-master节点创建资源清单：

```
[root@k8s-master ~]# vim labels.yaml
apiVersion: apps/v1
kind: Deployment
metadata:
  name: nginx-labels
  namespace: dean
spec:
  replicas: 3
  selector:
    matchLabels:
      web: nginx
  template:
    metadata:
      labels:
        web: nginx
    spec:
      containers:
        - name: nginx
          image: nginx:1.21
          imagePullPolicy: IfNotPresent
          ports:
            - containerPort: 80
```

02 应用资源：

```
[root@k8s-master ~]# kubectl apply -f labels.yaml
deployment.apps/nginx-labels created
```

该命令使用Kubectl工具将labels.yaml文件中定义的资源应用到Kubernetes集群中。使用-f选项指定要应用的文件名。

03 查看Pod标签：

```
[root@k8s-master ~]# kubectl get pods -n dean --show-labels
```

```
NAME                              READY   STATUS    RESTARTS   AGE    LABELS
    nginx-labels-89b74fcd9-96rxk    1/1     Running   0          24s
pod-template-hash=89b74fcd9,web=nginx
    nginx-labels-89b74fcd9-sgp8s    1/1     Running   0          24s
pod-template-hash=89b74fcd9,web=nginx
    nginx-labels-89b74fcd9-vjsqt    1/1     Running   0          24s
pod-template-hash=89b74fcd9,web=nginx
```

04 通过标签过滤来查看Pod：

```
[root@k8s-master ~]# kubectl get pods -n dean -l web=nginx
NAME                           READY   STATUS    RESTARTS   AGE
nginx-labels-89b74fcd9-96rxk   1/1     Running   0          2m31s
nginx-labels-89b74fcd9-sgp8s   1/1     Running   0          2m31s
nginx-labels-89b74fcd9-vjsqt   1/1     Running   0          2m31s
```

05 给一个Pod添加一个新的Labels：

```
[root@k8s-master ~]# kubectl label pods -n dean nginx-labels-89b74fcd9-96rxk auth=dean --overwrite
pod/nginx-labels-89b74fcd9-96rxk labeled

# -overwrite=false：如果为true，则允许覆盖标签，否则拒绝覆盖现有标签的更新标签
# 如果未指定或设置为false，则会报错：error: 'auth' already has a value (dean), and --overwrite is false
```

参数解析：

- kubectl：Kubernetes的命令行工具，用于与Kubernetes集群进行交互。
- label：Kubectl的一个子命令，用于给资源（如Pods、Deployments、Services等）添加或更新标签。
- pods：指定要操作的资源类型是Pod。
- -n dean：-n或--namespace参数的简写，指定要操作的命名空间是dean。命名空间用于在逻辑上将集群中的资源分组。
- nginx-labels-89b74fcd9-96rxk：指定要添加或更新标签的Pod的名称。这个名称通常是Pod的唯一标识符，由Deployment控制器根据Deployment名称和Pod模板的哈希值生成。
- auth=dean：指定要添加的标签，其中auth是键（key），dean是值（value）。标签是附加到资源的键-值对，可用于组织和选择资源。
- --overwrite：如果Pod上已经存在auth这个标签，该参数会确保旧的值被新值覆盖。没有这个参数，如果标签已存在，命令将失败。

06 通过新的标签来查看Pod：

```
[root@k8s-master ~]# kubectl get pods -n dean -l auth=dean
NAME                           READY   STATUS    RESTARTS   AGE
nginx-labels-89b74fcd9-96rxk   1/1     Running   0          6m23s
```

07 给default命名空间的一个Pod打上标签：

```
[root@k8s-master ~]# kubectl label pods deancurl auth=deanit --overwrite
pod/deancurl labeled
```

08 查看标签的键为auth，值中有deanit和dean的Pod：

```
[root@k8s-master ~]# kubectl get pods -A -l 'auth in (deanit,dean)'
NAMESPACE   NAME                          READY   STATUS    RESTARTS   AGE
default     deancurl                      1/1     Running   0          15d
dean        nginx-labels-89b74fcd9-96rxk  1/1     Running   0          16m
```

09 查看标签是web=nginx且auth不等于dean的Pod：

```
[root@k8s-master ~]# kubectl get pods -A -l web=nginx,auth!=dean
NAMESPACE   NAME                          READY   STATUS    RESTARTS   AGE
dean        nginx-labels-89b74fcd9-sgp8s  1/1     Running   0          22m
dean        nginx-labels-89b74fcd9-vjsqt  1/1     Running   0          22m
```

10 取消Pod的auth的标签：

```
root@k8s-master ~]# kubectl label pods -n dean nginx-labels-89b74fcd9-96rxk auth-
# 输出信息
pod/nginx-labels-89b74fcd9-96rxk labeled
```

该命令在Kubernetes集群中，使用kubectl label命令通过指定-（减号）作为要设置的值来"删除"该标签。不过，更准确地说，是通过不指定值并加上--overwrite标志来删除标签，因为-（减号）实际上并不是一个删除标签的直接操作，但Kubernetes的Kubectl工具在这种情况下会将其解释为删除标签。

11 再次查看auth标签的Pod，发现已经不存在了：

```
[root@k8s-master ~]# kubectl get pods -n dean -l auth=dean
No resources found in dean namespace.
```

运维笔记：如何在Kubernetes中使用标签进行资源管理和查询是运维人员的基本技能，本实验有助于我们加深对Kubernetes资源管理和查询操作的理解，提升实际应用能力。

3.2 InitC

在Kubernetes中，Init Container（通常缩写为InitC，即初始化容器）扮演着至关重要的角色。它作为Pod中的特殊容器，负责在主应用容器启动之前执行关键的配置和依赖性检查任务。本节将介绍Init Container的概念、特性、常见配置及其在实际场景中的应用，并通过实验加深对其的理解。

3.2.1 InitC 概述

在Kubernetes中，Init Container是一个核心概念。在一个Pod内，Init Container会在应用容器启动之前按顺序执行。它们通常用于执行初始化任务，比如配置环境、等待依赖服务准备就绪等。

Init Container作为Pod中主容器启动前运行的容器，负责执行一些初始化工作，包括配置文件和数据的下载、依赖服务的检查等。其主要目的是确保Pod中的主容器能够在一个正确且可靠的环境下启动和运行。

Init Container具有以下特点。

- 顺序执行：Pod中的Init Container按照定义的顺序依次执行，只有当前一个Init Container成功退出（即返回状态码为0）后，下一个才会开始执行。
- 独立生命周期：Init Container的生命周期与Pod中的主容器是独立的。即使主容器已经启动，Init Container仍然可以运行。
- 共享卷：Init Container可以与主容器共享相同的文件系统卷（Volume），这使得它们能够访问相同的文件或配置信息。
- 重启策略：如果Init Container失败（即返回非零状态码），Kubernetes会根据Pod的重启策略来决定是否重启Pod。默认情况下，如果Init Container失败，Kubernetes会不断重启Pod，直到Init Container成功为止。但如果Pod的restartPolicy设置为Never，则Pod不会重新启动。

Init Container通常在以下场景使用。

- 配置文件和数据下载：Init Container可以在主容器启动之前从外部源（如S3、Git仓库等）下载配置文件或数据，并将其保存到共享的文件系统卷中，供主容器使用。
- 依赖服务检查：Init Container可以检查Pod所依赖的其他服务（如数据库、消息队列等）是否可用。如果依赖服务不可用，Init Container可以等待或尝试重新连接，直到依赖服务可用为止。
- 环境初始化：Init Container可以执行一些环境初始化任务，如创建必要的目录、设置文件权限等。

3.2.2 InitC 实验

本实验用于测试Init Containers（初始化容器）的特性。我们将主要介绍如何配置并使用Init Containers在应用容器启动之前执行初始化任务，如设置环境变量、加载配置文件或依赖服务启动检测等。

实验步骤如下：

01 在k8s-master节点创建资源清单：

```yaml
[root@k8s-master ~]# vim InitContainer.yaml
---
# 创建一个名为myservice的service
apiVersion: v1
kind: Service
metadata:
  name: myservice
spec:
  ports:
  - protocol: TCP
    # service的端口
    port: 80
    # 这是容器内部应用实际监听的端口。当访问Service的80端口时，流量实际上会被转发到后端服务容器的9376端口
    targetPort: 9376
---
# 创建一个名为mydb的service
apiVersion: v1
kind: Service
metadata:
  name: mydb
spec:
  ports:
    - protocol: TCP
      port: 80
      targetPort: 9377
---
apiVersion: v1
kind: Pod
metadata:
  name: deanapp
  labels:
    app: deanapp
    version: v1
spec:
  # 又定义一组初始化容器InitC，然后定义两个不同的初始化容器
  initContainers:
    - name: init-myservice
      image: busybox
      # 不断尝试解析myservice这个DNS名称，直到成功为止（每2秒尝试一次，并在每次尝试之间打印一条消息）
      command: ['sh','-c','until nslookup myservice; do echo waiting for myservice; sleep 2; done;']
    - name: initmydb
      image: busybox
      command: ['sh','-c','until nslookup mydb; do echo waiting for mydb; sleep 2; done;']
  containers:
    # 创建一个名为myapp-container的容器，这个容器启动后会输出The app is running!内容，然后休眠3600秒（1小时），在此期间，容器不会退出
```

```
    - name: myapp-container
      image: busybox
      command: ['sh','-c','echo The app is running! && sleep 3600']
```

02 应用资源：

```
[root@k8s-master ~]# kubectl apply -f InitContainer.yaml
service/myservice created
service/mydb created
pod/deanapp created
```

该命令使用Kubectl工具将InitContainer.yaml文件中定义的资源应用到Kubernetes集群中。使用-f选项指定要应用的文件名。从输出信息可以看出，已经成功创建了以下资源：

- service/myservice created
- service/mydb created
- pod/deanapp created

03 查看Pod状态，以确认初始化完成：

```
--- 以下为Pod的更改状态，从初始化到Pod创建，再到运行
[root@k8s-master ~]# kubectl get pods
deanapp            0/1      Init:1/2           0         9m51s
...
deanapp            0/1      PodInitializing    0         13m
...
deanapp            1/1      Running            0         14m
```

通过本实验，我们不仅学习了如何配置Init Containers来确保主应用容器启动前的环境准备，还实践了如何使用Kubernetes的声明式语法来描述和管理复杂的应用部署。

运维笔记：在本实验中对Kubernetes对象的创建和状态检查，有助于加深我们对Kubernetes操作流程和对象生命周期管理的认识。

3.2.3 部署 Elasticsearch 服务时配置 InitC

在Kubernetes部署Elasticsearch服务时，我们需要确保内核参数vm.max_map_count大于262 144。这是因为Elasticsearch运行时要求这个参数必须大于262 144。为了实现这一点，我们可以使用InitContainer来修改内核参数。但是，这要求Kublet启动时必须添加--allow-privileged参数。然而，一般生产中不会添加这个参数，因此最好在系统供给时要求这个参数修改完成。

以下是一个简单示例，用于展示如何在Kubernetes部署Elasticsearch服务时配置InitC：

```
spec:
  initContainers:
    - name: init-sysctl
      image: busybox
```

```yaml
        imagePullPolicy: IfNotPresent
        command:
          - sysctl
          - -w
          - vm.max_map_count=262144
        securityContext:
          privileged: true
      containers:
      - image: docker.elastic.co/elasticsearch/elasticsearch:6.4.0
        name: es-data
        # 资源
        resources:
          # 最大限制,与资源requests不同的是,资源limits不受节点可分配资源量的约束,所有limits的总和允许超过节点资源总量的100%
          limits:
            cpu: 300m
            memory: 512Mi
          # 通过设置资源requests,我们指定了Pod对资源需求的最小值。调度器在将Pod调度到节点的过程中会用到该信息
          requests:
            cpu: 200m
            memory: 256Mi
        # 环境变量
        env:
        - name: network.host
          value: "_site_"
        - name: node.name
          value: "${HOSTNAME}"
        - name: discovery.zen.ping.unicast.hosts
          value: "es-cluster"
        - name: discovery.zen.minimum_master_nodes
          value: "2"
        - name: cluster.name
          value: "test-cluster"
        - name: node.master
          value: "false"
        - name: node.data
          value: "true"
        - name: node.ingest
          value: "false"
        - name: ES_JAVA_OPTS
          value: "-Xms128m -Xmx128m"
        volumeMounts:
        - name: es-cluster-storage
          mountPath: /usr/share/elasticsearch/data
      volumes:
      - name: es-cluster-storage
        emptyDir: {}
```

在这个示例中,我们创建了一个名为init-sysctl的InitContainer,它使用了BusyBox镜像,并通

过sysctl命令设置了vm.max_map_count参数。我们还为Elasticsearch容器设置了资源限制和请求、环境变量以及卷挂载，通过这些设置可以确保Elasticsearch正常运行。

3.3 容器钩子

容器钩子（Container Hooks）在容器的生命周期关键时刻发挥作用，允许执行自定义的操作，从而满足多样化的应用需求。本节将深入探讨Kubernetes中容器钩子的概念、类型以及在实际场景中的应用，并通过实验加深对这一重要特性的理解。

3.3.1 容器钩子概述

K8s（Kubernetes）的容器钩子是一个重要概念，用于在容器生命周期的特定阶段执行自定义操作。这些钩子允许容器在启动前和停止前执行特定的任务，以满足不同的应用需求。

1. 容器钩子的定义

容器钩子是指容器在Pod运行过程中的触发事件，它们可以用来执行一些容器启动前和容器停止前的准备操作。

在Kubernetes中，有两种主要的容器钩子：PostStart和PreStop。

2. PostStart钩子

- 执行时机：在容器创建之后立即执行，但在容器入口点之前执行的时间点是不确定的。
- 使用场景：适用于需要在容器启动后进行一些初始化操作的场景，如建立数据库连接、从外部下载文件等。
- 实现方式：可以通过Exec或HTTP两种方式实现。Exec方式允许执行一个特定的命令，而HTTP方式则是对容器上的特定端点执行HTTP请求。

3. PreStop钩子

- 执行时机：在容器终止之前是否立即调用此钩子，取决于API的请求或者管理事件，如活动探针故障、资源抢占、资源竞争等。
- 行为特性：它是阻塞的且同步的，必须在删除容器的调用之前完成。如果容器已经完全处于终止或完成状态，则对PreStop钩子的调用将失败。
- 使用场景：适用于需要在容器停止前执行一些清理操作或给进程一个清理数据的时间，以保证服务的请求正常结束。
- 实现方式：同样可以通过Exec或HTTP两种方式实现。

4. 钩子处理程序的执行

当生命周期管理钩子被调用时，Kubernetes 会在注册的容器中执行处理程序。钩子处理程序的调用是同步的，对于 PostStart 钩子，容器入口点和钩子异步触发，但如果钩子运行时间过长，可能影响容器达到 running 状态。

3.3.2 容器钩子实验

本实验用于测试 Kubernetes 中的容器生命周期钩子的特性，特别是容器的启动前（PostStart）和停止前（PreStop）钩子。通过这一实验，读者将掌握如何配置这些钩子来执行自定义操作，如资源初始化、日志记录、清理任务或优雅关闭服务等，从而增强对 Kubernetes 容器管理机制的理解与应用能力。

具体实验操作步骤如下：

01 在 k8s-master 节点创建资源清单：

```yaml
[root@k8s-master ~]# vim postStartStop.yaml
apiVersion: v1
kind: Pod
metadata:
  name: post-start-stop
  namespace: dean
spec:
  containers:
    - name: post-start-stop-container
      image: nginx:1.19
      imagePullPolicy: IfNotPresent
      # 生命周期
      lifecycle:
        # 启动探针，容器启动时先执行的命令
        postStart:
          exec:
            command: ["/bin/sh","-c","echo Hello from the postStart handler > /usr/share/message"]
        # 停止探针，容器停止前执行的命令
        preStop:
          exec:
            command: ["/bin/sh","-c","echo Hello from the postStop handler > /usr/share/message"]
```

02 我们在 postStartStop.yaml 文件中定义了一个 Pod，该 Pod 运行基于 nginx:1.19 镜像的容器。通过配置 Pod 的生命周期管理（lifecycle），我们设置了以下两个钩子。

- PostStart 钩子：在容器启动后立即执行，将 Hello from the postStart handler 写入 /usr/share/message 文件中。这个操作验证了容器启动后可以成功执行自定义脚本或命令。

- PreStop钩子：在容器终止前执行，将Hello from the postStop handler也写入同一个文件。由于Pod一旦结束就无法再次访问，因此无法直接查看由PreStop钩子写入的内容。这提示我们在实际生产环境中，应该将这类信息保存到持久化存储中，以便事后查看或调试。

03 应用资源：

```
[root@k8s-master ~]# kubectl apply -f postStartStop.yaml
pod/post-start-stop created
```

该命令使用Kubectl工具将postStartStop.yaml文件中定义的资源应用到Kubernetes集群中，成功创建了Pod资源。使用-f选项指定要应用的文件名。

04 使用kubectl get pods -n dean命令查看创建的Pod状态：

```
[root@k8s-master ~]# kubectl get pods -n dean
NAME                    READY   STATUS    RESTARTS   AGE
post-start-stop         1/1     Running   0          16m
```

可以看到，Pod处于正常运行状态。

05 使用kubectl exec -n dean post-start-stop -it -- cat /usr/share/message命令进入Pod内部，查看PostStart钩子执行的结果。

```
[root@k8s-master ~]# kubectl exec -n dean post-start-stop -it -- cat /usr/share/message
Hello from the postStart handler
```

参数解析：

- kubectl exec：主命令，在Kubernetes集群中的某个Pod内的容器中执行一个命令。
- -n dean：这是--namespace的缩写，表示想在名为dean的命名空间中执行这个命令。
- post-start-stop：这通常是Pod的名称或者正在尝试连接的容器的名称。但通常我们会指定Pod名称和一个特定的容器名称（如果Pod中有多个容器）。在这里，如果post-start-stop是一个Pod名称，并且该Pod中只有一个容器，那么这个命令会默认在这个容器上执行。但如果有多个容器，则需要明确指定容器名称，如kubectl exec -n dean post-start-stop -c <container-name> -it -- cat /usr/share/message。
- -it：这是以下两个选项的组合。
  - -i：--stdin=true的缩写，表示想保持STDIN打开，这样可以与容器中的命令进行交互（尽管在这个cat命令的上下文中，交互性可能不是必需的）。
  - -t：--tty=true的缩写，表示想为该命令分配一个伪终端（TTY），这在与需要用户交互的命令（如bash）一起使用时非常有用。
- --：这是一个分隔符，用于区分kubectl命令的参数和希望在容器中执行的命令的参数。这是为了确保在容器内部执行的命令参数不会被误解为Kubectl的参数。

- cat /usr/share/message：这是想在容器内部执行的命令。这里正在使用cat命令来查看/usr/share/message文件的内容。

上述命令告诉我们如何在Kubernetes集群内对Pod容器进行操作并检查容器内部文件系统的状态。需要注意的是，由于一旦Pod的生命周期结束，我们就无法再进入该Pod，从而无法直接观察到停止钩子所写入的数据。因此，在现实生产环境中，最佳实践是把启动和停止钩子生成的文件保存到Pod的持久化存储中，例如节点目录或NFS共享存储。这样的处理确保了即使在Pod生命周期结束后，这些关键的信息仍能被保留和访问。

本实验的要点：容器钩子允许在容器生命周期的关键阶段执行用户定义的动作，这增强了对容器行为的控制。

PostStart和PreStop钩子提供了在容器启动后和终止前执行命令的能力，这对于初始化、清理和故障排除等操作非常有用。在实际应用中，应考虑如何持久化钩子操作的结果，尤其是在容器销毁后需要保留的数据。

通过本实验，我们不仅了解了容器钩子的设置和应用，还实践了如何使用Kubectl进行资源的部署和管理，以及如何检查容器内部的操作结果。这些技能对于管理和调试Kubernetes中的容器至关重要。

3.4 探针

探针是Kubelet对容器执行的定期健康检查，旨在确保容器按照预期工作，并在出现问题时立即采取措施。为了执行这些检查，Kubelet会调用容器内部的特定处理程序。本节将详细介绍探针的类型、探测方式，并通过实验展示如何配置及使用启动探针（Startup Probe）来提高应用的稳定性和可靠性。

3.4.1 探针概述

探针是由Kubelet对容器执行的定期诊断，要执行诊断，可以使用Kubelet调用由容器实现的Handler。Kubelet通过以下三种类型的处理程序来执行对容器的健康检查。

- ExecAction：在容器内执行指定命令，如果命令退出时返回码为0，则认为诊断成功。
- TCPSocketAction：对指定端口上的容器的IP地址进行TCP检查，如果端口打开，则诊断被认为是成功的。
- HTTPGetAction：对指定端口和路径上的容器的IP地址执行HTTP Get请求，如果响应的状态码大于或等于200且小于400，则诊断被认为是成功的（2xx代表正常，3xx代表跳转，大于4xx，比如401、403、404、500、501，这些均为不正常）。

探测方式如下。

- **StartupProbe**：K8s1.16版本后新加的探测方式，用于判断容器内的应用程序是否已经启动。如果配置了startupProbe，就会先禁止其他的探测，直到它成功为止，成功后将不再进行探测。
- **livenessProbe（存活探测）**：指定容器是否正在运行，如果存活探测失败，则Kubelet会杀死容器，并且容器将受到其重启策略的影响，如果容器不提供存活探针，则默认状态为Success。
- **readinessProbe（就绪探测）**：指示容器是否准备好服务请求，如果就绪探测失败，端点控制器将从与Pod匹配的所有Service的端点中删除该Pod的IP地址，初始延迟之前的就绪状态默认为Failure，如果容器不提供就绪探针，则默认状态为Success。

探针有以下类型。

- **httpGet**：对Pod探针的内部发起HTTP请求，通过相应的返回码判断是否达到就绪状态。
- **exec**：在Pod内部执行命令，若命令执行结果返回为0，则表示达到就绪。
- **tcp Socket**：在Pod内部尝试建立TCP连接，若创建成功，则表示达到就绪状态。

3.4.2 StartUp Probe 启动探针实验

本实验用于测试Kubernetes中的启动探针（Startup Probe）功能，特别是利用TCP Socket作为健康检查手段。通过本实验将学会如何配置TCP Socket类型的启动探针来监控容器在启动过程中的健康状态，确保容器完全启动并准备好接收流量之前，不会被Kubernetes视作不健康而重启或终止，从而增强应用部署的稳定性和可靠性。

实验步骤如下：

01 在k8s-master节点上创建资源清单文件：

```
[root@k8s-master ~]# vim startupprobe-tcpsocket.yaml
apiVersion: v1
kind: Pod
metadata:
  namespace: dean
  name: startupprobe-tcp-pod
spec:
  containers:
    - name: startupprobe-tcp-container
      image: nginx:1.19
      imagePullPolicy: IfNotPresent
      ports:
        - containerPort: 80
          name: nginx-port
      startupProbe:
        # 通过tcpSocket类型进行探测
        tcpSocket:
```

```
            port: 81                # 此处将端口设置为81,目的是测试探测端口失败的效果
            initialDelaySeconds: 1  # 初始化时间,创建完Pod后多少秒后开始进行探测
            periodSeconds: 3        # 检测间隔
            timeoutSeconds: 3       # 超时时间
            successThreshold: 1     # 成功一次则为成功
            failureThreshold: 3     # 失败3次则为失败
```

02 应用资源:

```
[root@k8s-master ~]# kubectl apply -f startupprobe-tcpsocket.yaml
pod/startupprobe-tcp-pod created
```

使用Kubectl工具将startupprobe-tcpsocket.yaml文件中定义的资源应用到Kubernetes集群中。使用-f选项指定要应用的文件名。

03 查看Pod状态,可以看到Pod已经被重启一次:

```
[root@k8s-master ~]# kubectl get pods -n dean startupprobe-tcp-pod
NAME                  READY   STATUS    RESTARTS   AGE
startupprobe-tcp-pod  0/1     Running   1          72s
```

04 查看Pod详细信息中的事件,可以看到探测81端口失败的事件记录,因为Nginx实际监听的是80端口,所以81端口的探测肯定会失败:

```
[root@k8s-master ~]# kubectl describe pods -n dean startupprobe-tcp-pod
...
Events:
  Type     Reason     Age                From               Message
  ----     ------     ----               ----               -------
  Normal   Scheduled  84s                default-scheduler  Successfully assigned
yztest/startupprobe-tcp-pod to k8s-node2
  Normal   Killing    20s                kubelet            Container
startupprobe-tcp-container failed startup probe, will be restarted
  Normal   Pulled     19s (x2 over 83s)  kubelet            Container image "nginx:1.19"
already present on machine
  Normal   Created    19s (x2 over 83s)  kubelet            Created container
startupprobe-tcp-container
  Normal   Started    19s (x2 over 82s)  kubelet            Started container
startupprobe-tcp-container
  Warning  Unhealthy  8s (x6 over 77s)   kubelet            Startup probe failed: dial
tcp 10.244.169.156:81: connect: connection refused
```

通过本实验,我们不仅可以深入了解Kubernetes的启动探针机制,还可以实践如何配置和使用TCP Socket类型的探针来确保容器应用的健康检查。

3.4.3 Readiness Probe 就绪探针实验

本实验用于测试Kubernetes中的就绪探针(Readiness Probe)功能,特别是采用httpGet请求作为健康检查机制。通过本实验,我们将深入了解如何配置httpGet类型的就绪探针来监测容器是否已

准备好接收流量。这将确保只有在容器内部服务完全可用时，才会将流量路由到该容器，从而提升应用服务的整体可用性和用户体验。

实验步骤如下：

01 在k8s-master节点创建资源清单文件：

```
[root@k8s-master ~]# vim readinessprobe-httpGet.yaml
apiVersion: v1
kind: Pod
metadata:
  name: readiness-httpget-pod
  namespace: dean
spec:
  containers:
    - name: readiness-httpget-container
      image: nginx:1.19
      imagePullPolicy: IfNotPresent
      readinessProbe:
        httpGet:
          port: 80
          path: /dean.html       # 写一个不存在的文件，制造错误
        initialDelaySeconds: 3   # 初始化时间
        periodSeconds: 3         # 检测间隔
```

在Kubernetes主节点上编写资源清单文件，其中定义了一个Nginx容器，并设置就绪探针通过httpGet方式进行健康检查。探针配置为监听80端口，并尝试访问一个不存在的路径（/dean.html），这是为了故意制造错误以观察探针的行为。

02 应用资源：

```
[root@k8s-master ~]# kubectl apply -f readinessprobe-httpGet.yaml
pod/readiness-httpget-pod created
```

使用Kubectl工具将readinessprobe-httpGet.yaml文件中定义的资源应用到Kubernetes集群中。使用-f选项指定要应用的文件名。

03 查看Pod状态，可以看到Pod已经被重启一次：

```
[root@k8s-master ~]# kubectl get pods -n dean readiness-httpget-pod
NAME                    READY   STATUS    RESTARTS   AGE
readiness-httpget-pod   0/1     Running   1          72s
```

04 查看Pod详细信息中的事件，发现事件显示状态码404，这是因为Pod中不存在我们要探测的路径文件，导致健康检查失败，Pod未能标记为就绪状态：

```
[root@k8s-master ~]# kubectl describe pods -n dean readiness-httpget-pod
...
Events:
```

```
    Type     Reason     Age                   From                Message
    ----     ------     ----                  ----                -------
    Normal   Scheduled  73s                   default-scheduler   Successfully assigned
yztest/readiness-httpget-pod to k8s-node3
    Normal   Pulled     72s                   kubelet             Container image "nginx:1.19"
already present on machine
    Normal   Created    72s                   kubelet             Created container
readiness-httpget-container
    Normal   Started    72s                   kubelet             Started container
readiness-httpget-container
    Warning  Unhealthy  7s (x22 over 70s)     kubelet             Readiness probe failed: HTTP
probe failed with statuscode: 404
```

本实验强调就绪探针在确保容器健康和准备就绪方面的重要性。通过合理配置就绪探针，可以有效避免因容器内部服务未完全就绪而导致的请求失败或服务不稳定问题。此外，实验也展示了在配置就绪探针时，必须确保探针的端点存在且能正确响应，以避免因配置错误导致的不必要的服务中断。

3.4.4 LivenessProbe 存活探针实验

本实验用于测试Kubernetes中的存活探针（Liveness Probe）功能，特别是采用执行（exec）命令作为健康检查的方式。通过本实验将学会如何配置exec命令类型的存活探针来定期执行命令，并根据命令的退出状态来判断容器是否仍然健康运行。这将帮助Kubernetes及时发现并重启因内部故障或外部因素导致无法正常工作的容器，保障应用的高可用性和稳定性。

实验步骤如下：

01 在k8s-master节点创建资源清单文件：

```
[root@k8s-master ~]# vim livenessprobe-exec.yaml
apiVersion: v1
kind: Pod
metadata:
  name: liveness-exec-pod
  namespace: dean
spec:
  containers:
    - name: liveness-exec-container
      image: busybox
      imagePullPolicy: IfNotPresent
      # 容器执行的命令，在容器创建后会创建一个文件/tmp/live，然后休眠30秒，删除这个文件，再休眠1
分钟，在创建容器后的一分钟内文件是存在的，然后30秒后文件删除了，我们检测不到这个文件，那么该容器就被杀死
了，因为Pod中只有一个容器，那么容器死掉意味着Pod也会被重启，继续执行这个流程，会一直重启
      command: ["/bin/sh","-c","touch /tmp/live; sleep 30; rm -rf /tmp/live; sleep 60"]
      # 探测类型为存活探测
      livenessProbe:
        # 探测方式选择为exec，在容器中执行命令
        exec:
```

```
        command: ["test","-e","/tmp/live"]
      initialDelaySeconds: 1
      periodSeconds: 3
```

该资源清单YAML文件中定义了一个名为liveness-exec-pod的Pod，其中包含一个名为liveness-exec-container的容器。容器使用BusyBox镜像，并配置了存活探针，该探针通过执行命令["test","-e","/tmp/live"]来检查容器的健康状态。

02 应用资源：

```
[root@k8s-master ~]# kubectl apply -f livenessprobe-exec.yaml
pod/liveness-exec-pod created
```

使用Kubectl工具将livenessprobe-exec.yaml文件中定义的资源应用到Kubernetes集群中。使用-f选项指定要应用的文件名。

03 查看Pod状态，可以看到Pod已经启动成功并运行，因为探测已经成功：

```
[root@k8s-master ~]# kubectl get pods -n dean
NAME                READY   STATUS    RESTARTS   AGE
liveness-exec-pod   1/1     Running   0          4s
```

04 我们持续观察Pod状态，可以看到Pod被重启了两次：

```
[root@k8s-master ~]# kubectl get pods -n dean
NAME                READY   STATUS    RESTARTS   AGE
liveness-exec-pod   1/1     Running   2          3m19s
```

本实验展示了如何配置和使用Kubernetes中的存活探针，以确保容器的健康运行和自动恢复能力。通过合理设置存活探针，可以有效提高应用的可靠性和稳定性。

3.5 本章小结

通过本章的学习，我们不仅掌握了Label、容器钩子、探针这些核心概念的基本原理和使用方法，还深刻体会到了它们在Kubernetes容器编排和资源管理中的重要作用。未来，我们将继续深入学习Kubernetes的更多高级特性和最佳实践，不断提升自己的技术水平和实战能力。通过本章的学习和实践，读者可以对Kubernetes的Pod控制器有了一个深入的了解，为后续章节的学习打下基础。

第 4 章

Service服务发现与负载均衡

在前面的章节中，我们探讨了Kubernetes中用于部署和运行应用的几种核心对象。Pod作为部署微服务应用的基础单元，而Deployment则为其增加了扩缩容、自愈和滚动更新等高级功能。然而，尽管Deployment具备这些优势，它仍未能解决一个关键问题：我们不能仅通过Pod的IP地址直接访问它们。为此，Kubernetes引入了Service对象，它能为一组动态变化的Pod提供稳定且可靠的网络访问能力。

本章将介绍Service对象在服务发现与负载均衡中的应用。

4.1 Service原理

首先，明确一下术语。本书中以大写字母出现的Service，指的是Kubernetes中用来为Pod提供稳定的网络服务的Service对象。就像Pod、ReplicaSet或Deployment，一个Kubernetes Service是指我们在部署文件中定义的API中的一个REST对象，最终需要POST到API Server。Service对象的缩写为svc。

我们可以将Service理解为具有固定的前端和动态的后端的中间层。所谓前端，主要由IP、DNS名称和端口组成，始终不变；而后端则主要由一系列的Pod构成，时常会发生变化。

其次，每一个Service都拥有固定的IP地址、固定的DNS名称以及固定的端口。最后，Service与Pod之间是通过Label和Label筛选器（selector）松耦合在一起的，动态选择将流量转发至哪些Pod。Deployment和Pod之间也是通过这种方式进行关联的，这种松耦合方式是Kubernetes具备足够的灵活性的关键。

随着Pod频繁进行扩容和缩容、发生故障以及滚动升级等操作，Service会动态更新其维护的相匹配的健康Pod列表。具体来说，其中的匹配关系是通过Label筛选器（selector）和名为Endpoint对象的结构共同完成的。每一个Service在被创建时都会得到一个关联的Endpoint对象。整个Endpoint对象其实就是一个动态的列表，其中包含集群中所有的匹配Service Label筛选器的健康Pod。

Kubernetes会不断地检查Service的Label筛选器和当前集群中健康Pod的列表。如果有新的能够匹配Label筛选器的Pod出现，它就会被加入Endpoint对象，而消失的Pod则会被剔除。也就是说，Endpoint对象始终是保持更新的。这时，当Service需要将流量转发到Pod时，就会到Endpoint对象中最新的Pod列表中进行查找。当要通过Service转发流量到Pod时，通常会先在集群内部DNS中查询Service的IP地址。流量被发送到该IP地址后，会被Service转发到其中一个Pod。不过，Kubernetes原生应用（知悉Kubernetes集群并且能够访问Kubernetes API的应用）可以直接查询Endpoint API，而无须查找DNS和使用Service的IP。

在Kubernetes（K8s）中，Endpoint是一个关键的核心对象，它扮演着连接Service和后端Pod的重要角色。Endpoint具有以下特性：

- Endpoint代表了Service后端的一组IP地址和端口号，用于将流量从Service引导到实际运行应用程序的Pod。
- Endpoint提供了对服务后端的抽象，允许用户在集群中动态地管理服务的网络终端。
- Endpoint用于将服务的网络地址与后端容器或节点上的实际服务进行关联，实现服务的发现和访问。
- 通过Endpoint，K8s可以实现服务的动态发现和负载均衡，将流量负载均衡到后端Pod上，提高服务的高可用性和水平扩展性。
- Endpoint是根据Service和Pod的标签选择器自动生成的，因此当Service或Pod的标签发生变化时，Endpoint会自动更新，实现动态的服务发现和负载均衡。
- Endpoint帮助Kubernetes控制平面自动管理服务的网络终结点，从而提高服务的可用性和性能。

4.2 ClusterIP

ClusterIP是Kubernetes中Service对象的一个IP地址，这是一个虚拟IP地址，专门用于访问Service后面的Pod提供的服务。本节将介绍ClusterIP的原理与使用，了解其在负责均衡中的应用。

4.2.1 ClusterIP 概述

在Kubernetes集群中，ClusterIP为Service对象提供了一个稳定的IP地址，使得集群内部的其他组件或者Pod能够通过这个IP地址访问Service所代理的后端Pods。

每个Service对象在创建时都会被分配一个ClusterIP，这个IP地址会在Kubernetes集群的内部DNS系统中注册，允许通过DNS名称查询到Service。这种方式大大简化了服务间的通信，因为内部服务的消费者不需要知道后端Pod的详细信息，通过Service的ClusterIP和端口即可进行通信。

ClusterIP是虚拟存在的，它并不代表任何实际的网络接口，因此无法从集群外部直接访问。这也意味着外部流量不能直接通过ClusterIP到达服务，需要通过其他方式如NodePort或LoadBalancer来暴露服务至集群外。

当请求通过ClusterIP发送到Service时，Kube-proxy负责将这些请求转发到后端的Pods。Kube-proxy支持多种模式，其中一种常见的模式是iptables代理模式。在该模式下，ClusterIP主要在每个Node节点使用iptables，将发向ClutserIP对应端口的数据转发到Kube-proxy，然后Kube-proxy自己内部实现有负载均衡的方法，并且可以查询到这个Service下对应Pod的地址和端口，进而把数据转发给对应的Pod的地址和端口。

运维笔记：ClusterIP的设计为Kubernetes集群内部的服务发现和通信提供了便利，但需要注意它只作用于集群内部。

4.2.2 ClusterIP 实验

本实验用于测试Kubernetes中ClusterIP类型的Service资源，通过直接访问Service的Cluster IP地址，将观察到Kubernetes默认采用的负载均衡机制——轮询（Round Robin）。本实验将展示如何配置ClusterIP类型的Service，并验证当多个Pod副本作为后端时，客户端请求如何被均匀地分配到这些Pod上，从而确保服务的高可用性和负载均衡效果。

具体实验步骤如下：

01 在k8s-master节点创建资源清单：

```
[root@k8s-master ~]# vim myipClusterIP.yaml
apiVersion: apps/v1
kind: Deployment
metadata:
  name: myip-clusterip
  namespace: dean
spec:
  replicas: 3
  selector:
    matchLabels:
      myip: clusterip
  template:
    metadata:
      labels:
        myip: clusterip
    spec:
      containers:
        - name: myip
          image: gaopengju/whats-my-ip:latest
          imagePullPolicy: IfNotPresent
          ports:
            - containerPort: 8080
---
```

```yaml
# Service类型的API版本
apiVersion: v1
# 类型为Service
kind: Service
metadata:
  name: myip-clusterip
  namespace: dean
spec:
  # Service类型
  type: ClusterIP
  # Label筛选器，要与上面Deployment的筛选器的键-值对保持一致
  selector:
    myip: clusterip
  # 端口信息
  ports:
    # Service的端口
    - port: 38080
      # 此端口的别名
      name: http
      # 转发到后端Pod的端口
      targetPort: 8080
```

首先定义了一个名为myip-clusterip的Deployment，它包含三个副本，每个副本运行一个容器，该容器基于gaopengju/whats-my-ip:latest镜像，监听8080端口。接着，定义了一个同名的Service资源myip-clusterip，类型为ClusterIP，它将请求转发到后端Pod的8080端口，而自身监听38080端口。

02 应用资源：

```
[root@k8s-master ~]# kubectl apply -f myipClusterIP.yaml
deployment.apps/myip-clusterip created
service/myip-clusterip created
```

使用Kubectl工具将myipClusterIP.yaml文件中定义的资源应用到Kubernetes集群中。使用-f选项指定要应用的文件名。从输出信息可以看出，已经成功创建了Deployment和Service资源：

- deployment.apps/myip-clusterip created
- service/myip-clusterip created

03 查看Pod名及IP地址和Service信息，确认它们都是正确创建配置的：

```
# 通过shell命令格式化输出Pod信息
[root@k8s-master ~]# kubectl get pods -n dean -o wide | grep myip-clusterip | awk '{print $1 " " $6}'
# 输出信息
myip-clusterip-7fb85dd5b7-8hjtz    10.244.169.149
myip-clusterip-7fb85dd5b7-pb2d2    10.244.36.88
myip-clusterip-7fb85dd5b7-qcn94    10.244.107.216

# svc信息
```

```
[root@k8s-master ~]# kubectl get svc -n dean
NAME              TYPE        CLUSTER-IP       EXTERNAL-IP   PORT(S)      AGE
myip-clusterip    ClusterIP   10.102.200.41    <none>        38080/TCP    51s
redis-svc         ClusterIP   10.104.5.170     <none>        6379/TCP     20d
```

04 访问ClusterIP测试Service，可以看出Service的默认负载为轮询模式：

```
[root@k8s-master ~]# curl 10.102.200.41:38080
HOSTNAME:myip-clusterip-7fb85dd5b7-qcn94 IP:10.244.107.216

[root@k8s-master ~]# curl 10.102.200.41:38080
HOSTNAME:myip-clusterip-7fb85dd5b7-pb2d2 IP:10.244.36.88

[root@k8s-master ~]# curl 10.102.200.41:38080
HOSTNAME:myip-clusterip-7fb85dd5b7-8hjtz IP:10.244.169.149
```

这里，通过多次执行curl 10.102.200.41:38080命令访问Service的ClusterIP地址，观察到输出结果中Pod的IP地址发生变化，验证了Kubernetes的轮询负载均衡机制。同时展示了ClusterIP Service能够将请求均匀地分配给后端的多个Pod副本，从而实现高可用性和负载均衡。

4.3 NodePort

NodePort是Kubernetes中一种强大的服务类型，它设计用于将集群内部的服务安全且有效地暴露给外部访问。本节将深入探讨NodePort的核心概念和工作方式，并通过实战介绍其应用。

4.3.1 NodePort 概述

NodePort是Kubernetes服务的一种类型，允许集群外部的客户端通过每个节点的静态端口访问服务。

在Kubernetes中，NodePort类型的服务是对外暴露服务的基本方式之一。它在每个节点上开放一个特定的端口，任何发送到该端口的流量都会被转发到对应的服务。例如，如果一个Web应用需要被外部用户访问，则可以配置该服务的type为NodePort并指定一个nodePort（例如30001），这样外部用户就可以通过访问http://节点IP:30001来访问该Web服务。

NodePort在Kubernetes中的工作方式是在每个节点上配置一个静态端口，默认的端口范围是30000~32767。当流量到达任一节点上的这个端口时，Kube-Proxy都会将流量转发给后端的Pod。这种机制利用了Kube-Proxy创建的iptables规则，这些规则会自动将到达NodePort的流量重定向到正确的后端Pod。

与其他服务类型相比，例如LoadBalancer和Ingress，NodePort是最基础也是最灵活的外部流量接入方式之一。LoadBalancer依赖于底层云平台提供的负载均衡器，而Ingress则更侧重于HTTP层的流量管理和路由。实际上，无论选择哪种方式，NodePort通常都会是其基础组件之一。

4.3.2 NodePort 实验

本实验旨在通过实际操作验证Kubernetes中NodePort类型Service的配置及功能，NodePort会在集群中的每个节点上监听一个静态分配的端口，使得外部流量能够直接通过该端口访问集群内部的服务。

实验步骤如下：

01 在k8s-master节点创建资源清单：

```yaml
[root@k8s-master ~]# vim myipNodePort.yaml
apiVersion: apps/v1
kind: Deployment
metadata:
  name: myip-nodeport
  namespace: dean
spec:
  replicas: 3
  selector:
    matchLabels:
      myip: nodeport
  template:
    metadata:
      labels:
        myip: nodeport
    spec:
      containers:
        - name: myip
          image: gaopengju/whats-my-ip:latest
          imagePullPolicy: IfNotPresent
          ports:
            - containerPort: 8080
---
# Service类型的API版本
apiVersion: v1
# 类型为Service
kind: Service
metadata:
  name: myip-nodeport
  namespace: dean
spec:
  # Service类型
  type: NodePort
  # Label筛选器，要与上面Deployment的筛选器的键一值对保持一致
  selector:
    myip: nodeport
  # 端口信息
  ports:
    # Service的端口
```

```
        - port: 38080
          # 此端口的别名
          name: http
          # 转发到后端Pod的端口
          targetPort: 8080
          # 指定NodePort的端口，NodePort使用的是静态端口范围（30000~32767），因此需要确保这些端
口在集群节点上没有被其他服务占用
          nodePort: 32519
```

02 应用资源：

```
[root@k8s-master ~]# kubectl apply -f myipNodePort.yaml
deployment.apps/myip-nodeport created
service/myip-nodeport created
```

该命令使用Kubectl工具将myipNodePort.yaml文件中定义的资源应用到Kubernetes集群中。使用-f选项指定要应用的文件名。从输出信息可以看出，已经成功创建了以下资源：

- deployment.apps/myip-nodeport created
- service/myip-nodeport created

03 查看Pod名及IP地址和Service信息：

```
# 通过shell命令格式化输出Pod信息
[root@k8s-master ~]# kubectl get pods -n dean -o wide | grep myip-nodeport | awk '{print $1" "" "$6}'
myip-nodeport-5586dc6587-nsq86    10.244.36.89
myip-nodeport-5586dc6587-pfrgq    10.244.107.217
myip-nodeport-5586dc6587-qsc7g    10.244.169.150

# svc信息
[root@k8s-master ~]# kubectl get svc -n dean
NAME              TYPE        CLUSTER-IP       EXTERNAL-IP   PORT(S)            AGE
myip-clusterip    ClusterIP   10.102.200.41    <none>        38080/TCP          52m
myip-nodeport     NodePort    10.110.139.91    <none>        38080:32519/TCP    8s
redis-svc         ClusterIP   10.104.5.170     <none>        6379/TCP           20d
```

参数解析：

- NAME：Service的名字，在同一命名空间下不能重复。
- TYPE：Service的类型，常用的有ClusterIP、NodePort和LoadBalancer。
- CLUSTER-IP：分配的集群IP地址。
- EXTERNAL-IP：外部IP地址，一般Service类型为LoadBalancer才会分配IP地址。
- PORT(S)：Service暴露的端口，使用类型为NodePort时，前面的端口为服务端口，后面的端口为节点上暴露的端口。
- AGE：Service的年龄。

04 访问集群各节点的IP测试Service，可以看出每个节点都会监听一个NodePort端口：

```
[root@k8s-master ~]# curl 192.168.1.10:32519
HOSTNAME:myip-nodeport-5586dc6587-nsq86 IP:10.244.36.89
[root@k8s-master ~]# curl 192.168.1.11:32519
HOSTNAME:myip-nodeport-5586dc6587-pfrgq IP:10.244.107.217
[root@k8s-master ~]# curl 192.168.1.12:32519
HOSTNAME:myip-nodeport-5586dc6587-qsc7g IP:10.244.169.150
[root@k8s-master ~]# curl 192.168.1.13:32519
HOSTNAME:myip-nodeport-5586dc6587-nsq86 IP:10.244.36.89
```

通过本实验，我们将了解到如何配置NodePort类型的Service，并验证在集群的每个节点上是否成功监听了指定的端口，以及如何通过任意节点的该端口访问后端Pod提供的服务，从而扩展服务的可访问性。

4.4 Headless Service

Headless Service（无头服务）在Kubernetes中提供了一种独特的服务访问方式，与传统服务类型相比，它没有Cluster IP，而是直接将Pod IP和DNS记录暴露给客户端。本节将详细介绍Headless Service的特性，并通过实战展示其实际应用。

4.4.1 Headless Service 概述

Headless Service是Kubernetes中的一种特殊类型的集群服务，其特点主要体现在以下几个方面。

- 无Cluster IP：Headless Service的一个显著特点是它不会被分配Cluster IP。在Kubernetes中，Service通常会被分配一个Cluster IP，使得集群内部的服务可以通过这个IP进行访问。然而，对于Headless Service来说，其spec.clusterIP被设置为None，因此在实际运行时不会被分配Cluster IP。
- 直接暴露Pod IP和DNS记录：由于没有Cluster IP，Headless Service会直接暴露所有Pod的IP和DNS记录。这使得客户端可以直接访问Pod的IP地址，并使用这些IP地址进行负载均衡。
- 自定义负载均衡策略：由于Headless Service没有使用默认的负载均衡策略（即通过服务转发连接到符合要求的任一Pod上），因此它可以实现自定义的负载均衡策略。客户端可以根据自身的需求选择合适的Pod进行访问。
- 结合StatefulSet使用：Headless Service通常与StatefulSet一起使用，用于部署有状态的应用场景。StatefulSet是Kubernetes中用于管理有状态应用的一种控制器，它可以为Pod提供稳定的网络标识和存储卷。通过与Headless Service结合使用，可以确保有状态应用在集群中的稳定运行和访问。

Headless Service支持自定义负载均衡策略，通常与StatefulSet一起用于部署有状态的应用场景。

4.4.2 Headless Service 实验

本实验用于测试Kubernetes中Service资源的Headless类型，该类型的Service不会分配ClusterIP，而是允许客户端直接通过DNS解析到后端Pod的IP地址。

实验步骤如下：

01 在k8s-master节点创建资源清单：

```
[root@k8s-master ~]# vim myapp-headless.yaml
apiVersion: apps/v1
kind: Deployment
metadata:
  name: myapp-headless
  namespace: dean
spec:
  replicas: 3
  selector:
    matchLabels:
      app: myapp
  template:
    metadata:
      labels:
        app: myapp
    spec:
      containers:
        - name: myapp
          image: nginx:1.21
          imagePullPolicy: IfNotPresent
          ports:
            - containerPort: 80
---
apiVersion: v1
kind: Service
metadata:
  name: myapp-headless
  namespace: dean
spec:
  selector:
    app: myapp
  # 集群IP设置为None，将不会给Service分配IP地址
  clusterIP: "None"
  ports:
    - port: 80
      targetPort: 80
```

运维笔记：一旦Service创建成功，其信息将被添加至集群的CoreDNS配置中。同时，Service会被赋予一个主机名，该主机名也会被同步更新到集群的CoreDNS中。

02 应用资源:

```
[root@k8s-master ~]# kubectl apply -f myapp-headless.yaml
deployment.apps/myapp-headless created
service/myapp-headless created
```

使用Kubectl工具将myapp-headless.yaml文件中定义的资源应用到Kubernetes集群中。使用-f选项指定要应用的文件名。从输出信息可以看出,已经成功创建了以下资源:

- deployment.apps/myapp-headless created
- service/myapp-headless created

03 查看Pod和Service信息:

```
--- Pod的IP地址信息
[root@k8s-master ~]# kubectl get pods -n dean -o wide | grep myapp-headless | awk '{print $1 "  " $6}'
# 输出信息
myapp-headless-69b98884cc-q9zs8     10.244.36.91
myapp-headless-69b98884cc-s72qm     10.244.169.152
myapp-headless-69b98884cc-tnthk     10.244.107.219

--- Service信息
[root@k8s-master ~]# kubectl get svc -n dean
NAME              TYPE         CLUSTER-IP       EXTERNAL-IP   PORT(S)            AGE
myapp-headless    ClusterIP    None             <none>        80/TCP             9m15s
myip-clusterip    ClusterIP    10.102.200.41    <none>        38080/TCP          98m
myip-nodeport     NodePort     10.110.139.91    <none>        38080:32519/TCP    45m
redis-svc         ClusterIP    10.104.5.170     <none>        6379/TCP           20d
```

参数解析:

- -o: -o或--output参数用于指定输出格式。
- wide: 是kubectl get命令的一种输出格式,它提供了比默认格式更多的列。具体来说,它通常包括与Pod相关的Node信息,或者与其他资源相关的其他详细信息。
- |: UNIX管道符号,用于将一个命令的输出作为另一个命令的输入。
- grep myapp-headless: 使用grep命令来筛选包含myapp-headless的行。这通常用于筛选Pod名称或标签中包含myapp-headless的Pod。
- awk: 使用awk命令来处理文本并输出格式化后的结果。
- '{print $1 " " $6}': 对于每一行输入,打印第1个字段(通常是Pod名称)和第6个字段(通常是Pod的IP地址)。字段之间用两个空格分隔。

04 查看kube-system命名空间下的coredns的Pod IP地址:

```
[root@k8s-master ~]# kubectl get pods -n kube-system -o wide | grep coredns
coredns-6d8c4cb4d-frw8v   1/1   Running   1   30d   10.244.107.194   k8s-node3   <none>           <none>
```

```
coredns-6d8c4cb4d-lnfsp   1/1   Running   1 (30d ago)   30d   10.244.36.67
k8s-node1   <none>           <none>
```

05 无头服务中没有Service的IP地址,但是仍然可以通过访问域名的方式来访问Pod。域名的格式如下:

svc的名称.Pod的命名空间.当前集群的域名(因为我们没有更改).coredns的IP地址

通过上述命令格式进行域名解析,可以看出域名被解析到了Pod的IP地址:

```
[root@k8s-master ~]# dig -t A myapp-headless.dean.svc.cluster.local. @10.244.107.194

; <<>> DiG 9.11.4-P2-RedHat-9.11.4-26.P2.el7_9.15 <<>> -t A
myapp-headless.dean.svc.cluster.local. @10.244.107.194
;; global options: +cmd
;; Got answer:
;; WARNING: .local is reserved for Multicast DNS
;; You are currently testing what happens when an mDNS query is leaked to DNS
;; ->>HEADER<<- opcode: QUERY, status: NOERROR, id: 48010
;; flags: qr aa rd; QUERY: 1, ANSWER: 3, AUTHORITY: 0, ADDITIONAL: 1
;; WARNING: recursion requested but not available

;; OPT PSEUDOSECTION:
; EDNS: version: 0, flags:; udp: 4096
;; QUESTION SECTION:
;myapp-headless.dean.svc.cluster.local. IN A

;; ANSWER SECTION:
myapp-headless.dean.svc.cluster.local. 30 IN A 10.244.107.219
myapp-headless.dean.svc.cluster.local. 30 IN A 10.244.36.91
myapp-headless.dean.svc.cluster.local. 30 IN A 10.244.169.152

;; Query time: 0 msec
;; SERVER: 10.244.107.194#53(10.244.107.194)
;; WHEN: 一 6月 03 18:13:04 CST 2024
;; MSG SIZE  rcvd: 225
```

通过本实验,我们应该了解如何配置Headless Service,并利用Kubernetes集群内置的CoreDNS服务进行DNS查询。使用dig工具,能够验证Headless Service的DNS记录,确保它能够正确解析到所有后端Pod的IP地址,从而支持基于Pod IP的直接访问,适用于需要直接通信到Pod的特定场景。

4.5 ExternalName

ExternalName Service类型在Kubernetes中提供了一种灵活的方式来访问集群外部的服务,使得集群内部的Pod可以无缝地通过Service名称与外部服务通信。本节将深入探讨ExternalName的特性并通过实战演示其实际应用。

4.5.1 ExternalName 概述

ExternalName是Kubernetes中的一种特殊类型的Service，它用于将Service直接映射到集群外部的服务，如DNS记录或负载均衡器的DNS名称。当使用ExternalName Service类型时，集群内的Pod可以使用该Service名称进行通信，就好像它是一个普通的Kubernetes Service一样，但实际上请求将通过集群外部的服务进行路由。它允许将Kubernetes Service映射到任何有效的DNS名称，这使得Pod能够方便地访问集群外部的服务，而无须配置复杂的网络路由或端口转发规则。

4.5.2 ExternalName 实验

本实验将在Kubernetes的dean命名空间中展开，旨在探索并验证跨命名空间通信的能力。实验的核心目标是展示即使Pod部署在不同的命名空间中，它们依然能够通过Service的名称（而非具体的IP地址）相互访问。这一功能依赖于Kubernetes的服务发现机制，确保了即使在不直接暴露Pod IP的情况下，不同命名空间中的Pod也能高效、安全地进行通信。

实验步骤如下：

01 在k8s-master节点创建资源清单：

```
[root@k8s-master ~]# vim myapp.yaml
apiVersion: apps/v1
kind: Deployment
metadata:
  name: myapp
  namespace: dean
spec:
  replicas: 1
  selector:
    matchLabels:
      app: myapp
  template:
    metadata:
      labels:
        app: myapp
    spec:
      containers:
      - name: myapp
        image: gaopengju/myapp:v1
        ports:
        - name: http
          containerPort: 80
---
apiVersion: v1
kind: Service
metadata:
  name: myapp-svc
```

```
  namespace: dean
spec:
  selector:
    app: myapp
  # None表示是无头服务
  clusterIP: None
  ports:
  # service的端口
  - port: 80
    # 容器的端口
    targetPort: 80
```

02 在k8s-master节点创建ExternalNames类型的Service资源清单：

```
[root@k8s-master ~]# vim ExternalNames.yaml
kind: Service
apiVersion: v1
metadata:
  name: myapp-externalnames-svc
spec:
  # Service类型为ExternalName
  type: ExternalName
  # 外部名字，这里可以是集群中的服务地址，也可以是外部的公网地址
  externalName: myapp-svc.dean.svc.cluster.local
```

03 在k8s-master节点创建client客户端的资源清单：

```
[root@k8s-master ~]# vim myclient.yaml
apiVersion: apps/v1
kind: Deployment
metadata:
  name: client
spec:
  replicas: 1
  selector:
    matchLabels:
      app: client
  template:
    metadata:
      labels:
        app: client
    spec:
      containers:
      - name: client
        image: gaopengju/ops-utils:ubuntu
        command: ["sh","-c","sleep 3600"]
```

04 应用以上资源清单：

```
[root@k8s-master ~]# kubectl apply -f myapp.yaml
[root@k8s-master ~]# kubectl apply -f ExternalNames.yaml
```

```
[root@k8s-master ~]# kubectl apply -f myclient.yaml
```

该命令使用Kubectl工具将myapp.yaml、ExternalNames.yaml、myclient.yaml文件中定义的资源应用到Kubernetes集群中。使用-f选项指定要应用的文件名。

05 查看上面创建的资源情况，确保服务创建并启动成功：

```
--- 查看dean命名空间下Pod和svc的详细情况
[root@k8s-master ~]# kubectl get -n dean pods,svc
NAME                            READY   STATUS    RESTARTS   AGE
pod/myapp-66cb64c475-7962g      1/1     Running   0          46m

NAME                  TYPE        CLUSTER-IP   EXTERNAL-IP   PORT(S)   AGE
service/myapp-svc     ClusterIP   None         <none>        80/TCP    30m

--- 查看default命名空间下Pod和svc的详细情况
[root@k8s-master ~]# kubectl get pods,svc
NAME                            READY   STATUS    RESTARTS   AGE
pod/client-5645cc59-7sks4       1/1     Running   0          39m

NAME                              TYPE           CLUSTER-IP   EXTERNAL-IP                         PORT(S)   AGE
service/kubernetes                ClusterIP      10.96.0.1    <none>                              443/TCP   31d
service/myapp-externalnames-svc   ExternalName   <none>       myapp-svc.dean.svc.cluster.local    <none>    39m
```

通过kubectl exec命令进入名为client的Pod中，访问上面创建的ExternalNames类型的Service，通过下面的测试结果，可以看到访问的IP为dean命名空间下的svc：

```
--- 进入Pod中
[root@k8s-master ~]# kubectl exec -it client-5645cc59-7sks4 -- bash
--- 通过ping测试验证是否为dean命名空间下的Pod的IP地址
root@client-5645cc59-7sks4:/# ping myapp-externalnames-svc
PING myapp-svc.dean.svc.cluster.local (10.244.169.153): 56 data bytes
64 bytes from 10.244.169.153: icmp_seq=0 ttl=62 time=0.263 ms
64 bytes from 10.244.169.153: icmp_seq=1 ttl=62 time=0.251 ms
^C--- myapp-svc.dean.svc.cluster.local ping statistics ---
5 packets transmitted, 5 packets received, 0% packet loss
round-trip min/avg/max/stddev = 0.246/0.262/0.288/0.000 ms
--- 通过curl请求Pod中的服务
root@client-5645cc59-7sks4:/# curl myapp-externalnames-svc/hostname.html
myapp-66cb64c475-7962g
--- 通过nslookup进行域名解析
root@client-5645cc59-7sks4:/# nslookup myapp-externalnames-svc
Server:         10.96.0.10
Address:        10.96.0.10#53

myapp-externalnames-svc.default.svc.cluster.local canonical name = myapp-svc.dean.svc.cluster.local.
```

```
Name:    myapp-svc.dean.svc.cluster.local
Address: 10.244.169.153
```

运维笔记：本实验有助于我们深入理解Kubernetes网络模型中的服务抽象以及命名空间之间通信的实现细节。

4.6 LoadBalancer

LoadBalancer（负载均衡器）在Kubernetes中扮演着重要的角色，这不仅增强了应用的高可用性和可靠性，还提供了一种高效的方法来管理外部流量进入集群。本节将探讨LoadBalancer的关键概念和如何指定服务IP来优化访问。

4.6.1 LoadBalancer 概述

LoadBalancer是一种专门设计用来分配网络或应用程序流量的设备或软件服务，它确保流量平均地分布在多个服务器上。这样不仅可以提升系统的整体响应速度，还能增强服务的可用性。

通过使用LoadBalancer，客户端可以无缝地通过一个中央点访问后端的多个Pod，如图4-1所示，展示了客户端通过LoadBalancer访问后端Pod的典型链路。

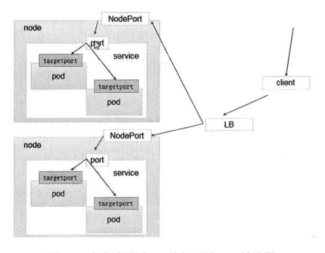

图 4-1　客户端通过 LB 访问后端 Pod 链路图

4.6.2 如何指定 LoadBalancer 类型的服务 IP

为了更精确地控制服务访问，Kubernetes允许在创建LoadBalancer类型的Service时手动指定IP地址。以下是一个YAML示例，展示了如何为名为nginx的Service指定一个loadBalancerIP：

```
apiVersion: v1
kind: Service
metadata:
```

```yaml
  name: nginx
  namespace: dean
  labels:
    app: nginx
spec:
  type: LoadBalancer
  # 若指定了IP，则使用此IP，若不指定，则使用自动分配的外部IP
  loadBalancerIP: xxx.xxx.xxx.xxx
```

在这个配置中，loadBalancerIP字段被设置为xxx.xxx.xxx.xxx，代表指定的IP地址。如果未指定，Kubernetes将自动分配一个外部IP地址给该Service。通过指定特定的IP地址，可以更灵活地管理网络流量和服务访问策略，特别是在多服务环境中。

4.7　Service端口范围及解除限制

在Kubernetes中，配置和管理Service端口范围是确保服务顺畅运行的关键一环。本节将详细介绍默认的Service端口范围，以及如何根据实际需求调整这一设置，从而优化集群的性能和资源利用。

4.7.1　Service 端口范围概述

在Kubernetes中，Service的端口范围及其解除限制的概念主要涉及NodePort类型的Service和Kubernetes集群的配置。

在Kubernetes中，NodePort类型的Service默认使用30000~32767的端口范围，这是为了防止端口冲突而设定的。这个范围适用于大多数情况，但如果特定场景需要更广或更具体的端口范围，Kubernetes也提供了自定义设置的能力。理解这一默认行为有助于更好地规划和部署服务，避免潜在的网络冲突。

下面将带领读者了解如何解除Service的端口限制。

4.7.2　Service 端口范围解除限制

对于需要修改默认端口范围的情况，通过编辑kube-apiserver的配置文件kube-apiserver.yaml，并通过设置--service-node-port-range参数，可以指定一个新的端口范围。

例如，在k8s-master节点执行修改APIServer配置文件：

```
vim /etc/kubernetes/manifests/kube-apiserver.yaml
```

将--service-node-port-range=10000-40000指定一个新的端口范围，如图4-2所示。

图 4-2 修改 Service 端口范围配置

上面的配置参数保存后,集群会监听到APIServer配置文件更改,Kubernetes集群会自动重启APIServer的Pod,当重启Pod时集群将不可用,需等待几分钟。等APIServer的Pod启动成功后,再次执行kubectl get pods命令,将看到集群返回查询结果:

```
[root@k8s-master ~]# kubectl get pods -n kube-system | grep apiserver
kube-apiserver-k8s-master                 1/1     Running   0          3m
```

运维笔记:了解并能够管理Kubernetes中Service的端口范围对于保障服务的稳定性与安全性至关重要。通过适当地配置和维护这些设置,管理员可以根据实际需求优化集群的网络性能和资源分配。

4.8 使用Service代理K8s外部应用

在Kubernetes集群中,有时我们需要访问外部应用或服务。为了实现这一目标,可以使用Service

代理功能将外部应用的域名解析到Kubernetes集群内部。本节将介绍如何使用Service代理K8s外部应用，并通过实验演示如何配置和使用这个功能。

4.8.1 使用 Service 代理 K8s 外部应用概述

在生产环境中，我们通常使用固定的名称（如域名）来访问外部应用或服务，而不是直接使用IP地址。然而，当某个项目正在迁移到Kubernetes集群时，部分服务可能仍然位于集群外部。在这种情况下，我们可以使用Service代理功能将外部服务的域名解析到Kubernetes集群内部，从而实现集群内部服务通过Service访问外部服务的目标。即使外部域名解析IP发生变化，我们也只需更改集群内部Endpoint的配置IP即可。

4.8.2 使用 Service 代理 K8s 外部应用实验

在本示例中，我们将带领读者了解如何将外部服务域名代理至Kubernetes集群内部，从而达到集群内部服务使用Service访问外部服务，即使外部域名解析IP更换，我们只需要更改集群内部Endpoint的配置IP即可。

具体操作步骤如下：

01 在k8s-master节点对域名进行ping操作以获取IP地址：

```
[root@k8s-master ~]# ping deanit.cn
PING deanit.cn (120.53.92.242) 56(84) bytes of data.
64 bytes from 120.53.92.242 (120.53.92.242): icmp_seq=1 ttl=52 time=8.33 ms
64 bytes from 120.53.92.242 (120.53.92.242): icmp_seq=2 ttl=52 time=8.10 ms
64 bytes from 120.53.92.242 (120.53.92.242): icmp_seq=3 ttl=52 time=7.72 ms
^C
--- deanit.cn ping statistics ---
3 packets transmitted, 3 received, 0% packet loss, time 2003ms
rtt min/avg/max/mdev = 7.729/8.053/8.330/0.268 ms
```

02 然后，在k8s-master节点上创建资源清单文件：

```
[root@k8s-master ~]# vim vim deanit-svc-external.yaml
apiVersion: v1
kind: Service
metadata:
  name: deanit-svc-external
  namespace: dean
  labels:
    app: deanit-svc-external
spec:
  ports:
  - name: http
    port: 80
    protocol: TCP
    targetPort: 80
```

```yaml
    - name: https
      port: 443
      protocol: TCP
      targetPort: 443
  sessionAffinity: None
  type: ClusterIP
---
apiVersion: v1
kind: Endpoints
metadata:
  # 这个要和Service名字一致,才能建立链接
  name: deanit-svc-external
  namespace: dean
  labels:
    app: deanit-svc-external
subsets:
- addresses:
  # 域名的解析IP
  - ip: 120.53.92.242
  ports:
  - name: http
    port: 80
    protocol: TCP
  - name: https
    port: 443
    protocol: TCP
```

我们创建了deanit-svc-external.yaml文件,并定义了Service和Endpoint资源。

03 最后,使用Kubectl工具将资源清单文件应用到Kubernetes集群中:

```
[root@k8s-master ~]# kubectl apply -f deanit-svc-external.yaml
service/deanit-svc-external created
endpoints/deanit-svc-external created
```

根据输出信息,可以看出已经成功创建了以下资源:

- service/deanit-svc-external created
- endpoints/deanit-svc-external created

04 查看Service和Endpoint资源:

```
[root@k8s-master ~]# kubectl get svc,ep -n dean | grep deanit-svc-external
service/deanit-svc-external   ClusterIP   10.105.23.32   <none>   80/TCP,443/TCP   3m39s
endpoints/deanit-svc-external   120.53.92.242:80,120.53.92.242:443   3m39s
```

实验结果验证了Service代理功能的有效性,使得集群内部服务可以通过Service访问外部服务。

运维笔记：本实验演示了如何在Kubernetes集群中使用Service代理功能来访问外部应用或服务。这种功能对于在迁移过程中保持服务的连续性和稳定性至关重要。同时，它也为Kubernetes集群提供了一种灵活的方式来管理外部资源的访问。

4.9 本章小结

通过本章的学习，读者将学习到Service如何作为Pods的逻辑集合的访问入口，通过为这些Pods提供稳定的IP地址和DNS名称，无论Pods的实际IP如何变化，Service都能确保服务的连续性。这一特性极大地增强了应用的可扩展性和容错性。其次，本章深入探讨了Service的几种类型，包括ClusterIP（集群内部访问）、NodePort（集群外部访问，通过每个节点的静态端口映射）、LoadBalancer（使用云提供商或外部地址提供负载均衡器对外暴露服务）以及ExternalName（通过CNAME记录将服务映射到外部DNS名称）。这些不同类型的Service为不同的部署需求提供了灵活的解决方案。此外，读者还将学习到如何通过Service的定义文件（YAML或JSON格式）来创建和管理Service，包括设置Selector以指定哪些Pods被包含在Service中，配置端口映射以满足不同的访问需求，以及使用标签和注解来增强Service的元数据管理。通过本章的学习和实践，读者可以对Kubernetes的Service有深入的理解，为后续章节的学习打下基础。

第 5 章

Ingress-Nginx服务网关

随着Kubernetes集群中微服务数量的增长,如何高效、安全地暴露这些服务给外部用户访问成为一个关键问题。本章将首先介绍Ingress-Nginx的基本概念,让读者对其有一个全面的认识。接着,详细讲解如何安装和配置Ingress-Nginx,并演示如何通过定义Ingress资源对象来创建和管理路由规则。此外,我们还将探讨一些高级配置和最佳实践,帮助读者优化Ingress-Nginx的性能和安全性。

5.1 Ingress-Nginx概述

Ingress-Nginx作为Kubernetes的一个扩展组件,不仅提供了基于域名、URL路径等条件的路由功能,还具备负载均衡、SSL终止、HTTP/2支持等高级特性。通过Ingress-Nginx,我们可以更灵活地管理集群的入口流量,提升服务的可访问性和安全性。

5.2 Ingress-Nginx安装

安装Ingress-Nginx组件的具体步骤如下:

01 首先,在k8s-master节点上创建一个资源清单文件。这个文件包含安装Ingress-Nginx组件所需的所有配置信息。以下是关键配置部分:

```
--- 资源清单内容过长,完整资源清单存放于本书附件内
--- 以下为清单关键配置
...
    spec:
      hostNetwork: true
      dnsPolicy: ClusterFirstWithHostNet
      containers:
```

```yaml
      - args:
        - /nginx-ingress-controller
        - --publish-service=$(POD_NAMESPACE)/ingress-nginx-controller
        - --election-id=ingress-controller-leader
        - --controller-class=k8s.io/ingress-nginx
        - --ingress-class=nginx
        - --configmap=$(POD_NAMESPACE)/ingress-nginx-controller
        - --validating-webhook=:8443
        - --validating-webhook-certificate=/usr/local/certificates/cert
        - --validating-webhook-key=/usr/local/certificates/key
        env:
        - name: POD_NAME
          valueFrom:
            fieldRef:
              fieldPath: metadata.name
        - name: POD_NAMESPACE
          valueFrom:
            fieldRef:
              fieldPath: metadata.namespace
        - name: LD_PRELOAD
          value: /usr/local/lib/libmimalloc.so
        image: docker.m.daocloud.io/heidaodageshiwo/controller:v1.3.0
        imagePullPolicy: IfNotPresent
...
```

02 接下来，使用kubectl apply -f Ingress-Nginx.yaml命令将资源清单文件中定义的资源应用到Kubernetes集群中。

```
[root@k8s-master ~]# kubectl apply -f Ingress-Nginx.yaml
namespace/ingress-nginx created
serviceaccount/ingress-nginx created
serviceaccount/ingress-nginx-admission created
role.rbac.authorization.k8s.io/ingress-nginx created
role.rbac.authorization.k8s.io/ingress-nginx-admission created
clusterrole.rbac.authorization.k8s.io/ingress-nginx created
clusterrole.rbac.authorization.k8s.io/ingress-nginx-admission created
rolebinding.rbac.authorization.k8s.io/ingress-nginx created
rolebinding.rbac.authorization.k8s.io/ingress-nginx-admission created
clusterrolebinding.rbac.authorization.k8s.io/ingress-nginx created
clusterrolebinding.rbac.authorization.k8s.io/ingress-nginx-admission created
configmap/ingress-nginx-controller created
service/ingress-nginx-controller created
service/ingress-nginx-controller-admission created
daemonset.apps/ingress-nginx-controller created
job.batch/ingress-nginx-admission-create created
job.batch/ingress-nginx-admission-patch created
ingressclass.networking.k8s.io/nginx created
validatingwebhookconfiguration.admissionregistration.k8s.io/ingress-nginx-admissi
on created
```

根据输出信息，可以看出已经成功创建包括命名空间、服务账户、角色、集群角色、角色绑定、集群角色绑定、配置映射、服务、守护进程集、作业等资源：

- namespace/ingress-nginx created
- serviceaccount/ingress-nginx created
- serviceaccount/ingress-nginx-admission created
- role.rbac.authorization.k8s.io/ingress-nginx created
- role.rbac.authorization.k8s.io/ingress-nginx-admission created
- clusterrole.rbac.authorization.k8s.io/ingress-nginx created
- clusterrole.rbac.authorization.k8s.io/ingress-nginx-admission created
- rolebinding.rbac.authorization.k8s.io/ingress-nginx created
- rolebinding.rbac.authorization.k8s.io/ingress-nginx-admission created
- clusterrolebinding.rbac.authorization.k8s.io/ingress-nginx created
- clusterrolebinding.rbac.authorization.k8s.io/ingress-nginx-admission created
- configmap/ingress-nginx-controller created
- service/ingress-nginx-controller created
- service/ingress-nginx-controller-admission created
- daemonset.apps/ingress-nginx-controller created
- job.batch/ingress-nginx-admission-create created
- job.batch/ingress-nginx-admission-patch created
- ingressclass.networking.k8s.io/nginx created
- validatingwebhookconfiguration.admissionregistration.k8s.io/ingress-nginx-admission created

03 最后，使用kubectl get pods,svc -n ingress-nginx命令查看Ingress-Nginx的Pod和Service的状态。从输出结果中，我们可以看到各个Pod都在正常运行，而服务也已经被正确创建：

```
[root@k8s-master ~]# kubectl get pods,svc -n ingress-nginx
NAME                                             READY   STATUS      RESTARTS   AGE
pod/ingress-nginx-admission-create-n244t         0/1     Completed   0          20m
pod/ingress-nginx-admission-patch-xzfdv          0/1     Completed   0          20m
pod/ingress-nginx-controller-f9sdr               1/1     Running     0          16m
pod/ingress-nginx-controller-rv7tx               1/1     Running     0          16m
pod/ingress-nginx-controller-tmwht               1/1     Running     0          16m

NAME            TYPE         CLUSTER-IP       EXTERNAL-IP   PORT(S)                      AGE
service/ingress-nginx-controller              NodePort    10.110.206.243   <none>        80:37042/TCP,443:24895/TCP   17m
service/ingress-nginx-controller-admission    ClusterIP   10.102.35.136    <none>        443/TCP                      20m
```

Pod解析：

- ingress-nginx-admission-create：Kubernetes中的准入控制器可以拦截请求（如Pod的创建请求），并根据某些规则进行修改或拒绝这些请求。
- ingress-nginx-admission-patch：在Kubernetes中，patch是一种更新资源的机制，它允许用户仅修改资源的部分字段，而不是整个资源。
- ingress-nginx-controller：Ingress Controller负责实现这些规则，并监听Ingress对象的更改以动态更新路由配置。

在k8s-master节点创建服务资源清单，用于测试Ingress-Nginx安装是否成功：

```
[root@k8s-master ~]# vim whats-my-ip.yaml
apiVersion: apps/v1
kind: Deployment
metadata:
  name: whats-my-ip
  namespace: dean
spec:
  selector:
    matchLabels:
      name: whats-my-ip
  replicas: 2
  template:
    metadata:
      labels:
        name: whats-my-ip
    spec:
      containers:
        - name: whats-my-ip
          image: gaopengju/whats-my-ip:latest
          imagePullPolicy: IfNotPresent
          ports:
            - containerPort: 8080
---
apiVersion: v1
kind: Service
metadata:
  name: whats-my-ip
  namespace: dean
spec:
  ports:
    - port: 8080
      targetPort: 8080
      protocol: TCP
  selector:
    name: whats-my-ip
---
```

```yaml
# api版本
apiVersion: networking.k8s.io/v1
kind: Ingress
metadata:
  annotations:
    # 我们这个接口用的是一个叫ingress.class为nginx的ingress，因为集群可能不止一个ingress
    kubernetes.io/ingress.class: "nginx"
  # ingress的名称
  name: myipweb
  namespace: dean
# 规格
spec:
  # 定义后端转发的规则，一个Ingress可以配置多个rules
  rules:
  # 通过域名进行转发
  - host: whats-my-ip.deanit.cn
    http:
      paths:
      # 配置后端服务
      - backend:
          service:
            name: whats-my-ip
            port:
              number: 8080
        path: /
        pathType: Prefix
```

这里，在k8s-master节点上创建了一个名为whats-my-ip的服务资源清单，用于测试Ingress-Nginx的安装是否成功。

04 应用资源：

```
[root@k8s-master ~]# kubectl apply -f whats-my-ip.yaml
deployment.apps/whats-my-ip created
service/whats-my-ip created
ingress.networking.k8s.io/myipweb created
```

使用Kubectl工具将whats-my-ip.yaml文件中定义的资源应用到Kubernetes集群中。根据输出信息，可以看出已经成功创建了Deployment、Service和Ingress资源。

05 查看Pod、Service和Ingress的状态：

```
--- 查看Pod资源信息
[root@k8s-master ~]# kubectl get pods -n dean
NAME                              READY   STATUS    RESTARTS   AGE
whats-my-ip-58848997bc-5fdc4      1/1     Running   0          7m5s
whats-my-ip-58848997bc-jt6gr      1/1     Running   0          7m5s

--- 查看Service资源信息
[root@k8s-master ~]# kubectl get svc -n dean
```

```
NAME            TYPE         CLUSTER-IP       EXTERNAL-IP    PORT(S)      AGE
whats-my-ip     ClusterIP    10.98.235.108    <none>         8080/TCP     7m11s

--- 查看Ingress资源信息
[root@k8s-master ~]# kubectl get ingress -n dean
NAME       CLASS    HOSTS                    ADDRESS          PORTS    AGE
myipweb    nginx    whats-my-ip.deanit.cn    10.110.206.243   80       7m15s
```

通过查看Pod、Service和Ingress的状态，确认它们都已正确创建。

06 添加hosts并访问测试：

```
--- 添加hosts用于域名和IP作为临时解析
[root@k8s-master ~]# echo "192.168.1.12    whats-my-ip.deanit.cn">>/etc/hosts

--- 访问域名进行测试，根据下面的输出结果可以得出访问域名已经到达Pod，并轮询访问各Pod
[root@k8s-master ~]# curl whats-my-ip.deanit.cn
HOSTNAME:whats-my-ip-58848997bc-5fdc4 IP:10.244.107.225

[root@k8s-master ~]# curl whats-my-ip.deanit.cn
HOSTNAME:whats-my-ip-58848997bc-jt6gr IP:10.244.36.99
```

通过添加hosts解析和访问域名进行测试，验证了Ingress配置的正确性以及轮询访问各Pod的功能。

5.3　Annotations注解

在Kubernetes中，Ingress Annotations是一组专门应用于Ingress对象的元数据。这些注解为Ingress控制器提供了附加的配置选项和命令，允许用户根据特定的需求对Ingress资源的行为进行定制。与ConfigMap全局生效不同，Annotations仅对指定的Ingress对象生效。

5.3.1　流量复制

我们有时需要将服务引流到测试环境进行测试，这时就需要使用Nginx的流量复制功能。

我们的需求是将www.deanit.cn/api/的流量复制到192.168.0.2的负载均衡上，另外，主机名与原主机一致。资源清单文件的配置如下：

```
apiVersion: networking.k8s.io/v1
kind: Ingress
metadata:
  name: web-mirror
  namespace: dean
  annotations:
    # 指定一个镜像主机，将请求镜像到这个主机
nginx.ingress.kubernetes.io/mirror-host: "www.deanit.cn"
    # 指定镜像目标URL,将请求镜像到这个URL
```

```
      nginx.ingress.kubernetes.io/mirror-target: http://192.168.0.2$request_uri
    spec:
      rules:
      - host: www.deanit.cn
        http: &http_rules
          paths:
          - backend:
              service:
                name: web-service
                port:
                  number: 80
            path: /api/
            pathType: ImplementationSpecific
```

5.3.2　IP 白名单

可以通过使用 nginx.ingress.kubernetes.io/whitelist-source-range 注解来设置 IP 白名单,以限制哪些 IP 地址可以访问你的服务。以下是不同场景的示例:

```
# 允许单个IP：只允许来自192.168.1.2的访问
nginx.ingress.kubernetes.io/whitelist-source-range: 192.168.1.2
# 允许一个C段：允许来自192.168.1.0/24网段内的所有IP访问
nginx.ingress.kubernetes.io/whitelist-source-range: 192.168.1.0/24
# 允许一个B段：允许来自192.168.0.0/16网段内的所有IP访问
nginx.ingress.kubernetes.io/whitelist-source-range: 192.168.0.0/16
# 允许多个IP源：允许来自192.168.0.0/16网段以及单个IP 192.168.1.2的访问
nginx.ingress.kubernetes.io/whitelist-source-range: 192.168.0.0/16,192.168.1.2
```

以下是一个 Ingress 资源清单的示例,用于将外部流量路由到名为 rabbitmq 的服务,并设置了 IP 白名单:

```
--- 资源清单如下
apiVersion: networking.k8s.io/v1
kind: Ingress
metadata:
  name: rabbitmq
  namespace: dean
  annotations:
    nginx.ingress.kubernetes.io/whitelist-source-range: 192.168.0.0/24,127.0.0.1
spec:
  rules:
  - host: rabbitmq.deanit.cn
    http: &http_rules
      paths:
      - backend:
          service:
            name: rabbitmq
            port:
              number: 15672
```

```
      path: /
      pathType: ImplementationSpecific
```

在这个示例中，Ingress资源将被配置为仅接受来自192.168.0.0/24网段和本地回环地址（127.0.0.1）的请求。所有其他来源的请求将收到403禁止响应。

5.3.3 IP黑名单

对单个域名进行限制，拒绝192.168.1.2的IP访问，允许其他IP访问：

```
apiVersion: networking.k8s.io/v1
kind: Ingress
metadata:
  name: rabbitmq
  namespace: dean
  annotations:
    nginx.ingress.kubernetes.io/server-snippet: deny 192.168.1.2;allow all;
spec:
  rules:
  - host: rabbitmq.deanit.cn
    http: &http_rules
      paths:
      - backend:
          service:
            name: rabbitmq
            port:
              number: 15672
        path: /
        pathType: ImplementationSpecific
```

配置说明：

- nginx.ingress.kubernetes.io/server-snippet：该注解允许直接写入Nginx配置片段。在这里，我们使用了deny 192.168.1.2;来拒绝特定IP的访问，然后用allow all;来允许所有其他IP的访问。
- spec.rules：定义了流量路由规则。这里的规则是针对rabbitmq.deanit.cn域名，将所有流量路由到名为rabbitmq的服务的15672端口。

在这个例子中，我们对域名rabbitmq.deanit.cn进行了设置，拒绝来自IP地址192.168.1.2的访问，同时允许其他所有IP地址的访问。通过这种方式可以精确地控制哪些IP地址可以访问你的服务，哪些不可以，从而增加了应用的安全性。

5.3.4 域名转发

由于第三方系统对域名的限制，我们需要对访问www.deanit.cn/api/weixin/的接口进行转发，使其指向api.deanit.cn/api/weixin/。为了实现这一需求，可以通过Kubernetes Ingress资源进行配置：

```yaml
apiVersion: networking.k8s.io/v1
kind: Ingress
metadata:
  name: www-api-weixin
  namespace: dean
  annotations:
    nginx.ingress.kubernetes.io/upstream-vhost: "api.deanit.cn"
spec:
  rules:
  - host: www.deanit.cn
    http: &http_rules
      paths:
      - backend:
          service:
            name: api-service
            port:
              number: 80
        path: /api/weixin/
        pathType: ImplementationSpecific
```

配置说明：

- nginx.ingress.kubernetes.io/upstream-vhost：该注解指定了转发的服务器域名，即api.deanit.cn。
- spec.rules：定义了转发规则。当访问www.deanit.cn的/api/weixin/路径时，请求将被转发到名为api-service的服务，端口为80。

通过上述配置，可以在不更改客户端代码的情况下，解决因第三方系统限制导致的接口访问问题。

5.3.5 返回字符串

例如，使用第三方平台的TXT文件验证时，就可以使用此注解来实现。

```yaml
apiVersion: networking.k8s.io/v1
kind: Ingress
metadata:
  name: face-geturl
  namespace: dean
  annotations:
    nginx.ingress.kubernetes.io/configuration-snippet: |
      #default_type application/json;              # 返回类型
      return 200 'Hello I am Dean';                # 直接返回字符
spec:
  rules:
  - host: www.deanit.cn
    http: &http_rules
      paths:
      - backend:
          service:
```

```
            name: web-service
            port:
              number: 80
        path: /deanVerify.txt
        pathType: ImplementationSpecific
```

配置说明：

- nginx.ingress.kubernetes.io/configuration-snippet：该注解允许在Nginx配置文件中插入自定义代码片段。在这里，我们使用了一个简单的返回语句来返回一个字符串"Hello I am Dean"作为响应内容。
- spec.rules：定义了转发规则。当访问www.deanit.cn的/deanVerify.txt路径时，请求将被转发到名为web-service的服务，端口为80。

通过上述配置可以实现对第三方平台的TXT文件验证，而无须修改客户端代码。

5.3.6 文件上传大小

文件上传大小限制可以通过修改ingress-nginx-controller的ConfigMap配置参数来实现。具体操作如下：

01 使用以下命令编辑ingress-nginx-controller的ConfigMap配置：

```
[root@k8s-master ~]# kubectl edit cm -n ingress-nginx ingress-nginx-controller
```

02 在打开的编辑器中，将以下参数配置到ConfigMap中即可：

```
data:
  allow-snippet-annotations: "true"        # 允许使用注解
  client_body_buffer_size: 10240m          # 客户端请求主体缓冲区大小
  client_max_body_size: 10240m             # 客户端请求主体最大大小
  proxy-body-size: 10240m                  # 代理主体大小
```

03 保存并退出编辑器。

此外，如果在某个Ingress配置中设置了相应的注解，那么这个设置会覆盖ingress-nginx-controller中的相应参数。例如：

```
annotations:
  nginx.ingress.kubernetes.io/proxy-body-size: "5m"
```

这里的nginx.ingress.kubernetes.io/proxy-body-size注解设置了代理主体大小为5MB。

5.3.7 域名HTTPS访问

通常在生产环境中，网站需要使用SSL证书，此示例将带领读者了解如何在Ingress-Nginx网关中配置证书。

首先使用如下命令从SSL证书厂商获取证书文件，例如本实验获取的证书名为deanit.key和deanit.crt：

```
[root@k8s-master ~]# kubectl create secret tls deanit-tls --key deanit.key --cert deanit.crt
```

参数解析：

- kubectl create secret tls：这是创建TLS类型Secret的命令。
- deanit-tls：这是创建的Secret的名称，它将在Kubernetes集群中唯一标识这个Secret。
- --key deanit.key：这个选项指定了私钥文件的路径。在这个例子中，私钥文件名为deanit.key。
- --cert deanit.crt：这个选项指定了证书文件的路径。在这个例子中，证书文件名为deanit.crt。

服务的Ingress资源清单配置如下：

```yaml
apiVersion: networking.k8s.io/v1
kind: Ingress
metadata:
  # Ingress名字
  name: deanit
spec:
  # tls配置
  tls:
    # 域名
    - hosts:
        - www.deanit.cn
    # 根据域名SSL证书创建的Secret
    - secretName: deanit-tls
  rules:
  - host: www.deanit.cn
    http:
      paths:
      - path: /
        backend:
          service:
            name: deanit
            port:
              number: 80
```

Ingress默认访问HTTPS，如果不想强制跳转到HTTPS，可以配置注解或者修改Ingress的ConfigMap配置参数。

默认情况下，如果为该入口启用了TLS，控制器会重定向（308）到HTTPS。如果要全局禁用此行为，可以在Ingress控制器修改ConfigMap。

注：Ingress控制器在升级后默认的永久重定向状态码从301变成了308。

```
kubectl edit cm -n ingress-nginx ingress-nginx-controller
# 添加此参数
ssl-redirect: "false"
```

如要实现单独域名的HTTPS不强制跳转，可配置以下注解来实现：

```
nginx.ingress.kubernetes.io/ssl-redirect: "false"
```

在集群外使用SSL卸载（例如AWS ELB）时，即使没有可用的TLS证书，强制重定向到HTTPS也可能很有用。这可以通过使用以下注解来对特定资源实现：

```
nginx.ingress.kubernetes.io/force-ssl-redirect: "true"
```

要保留URI中的尾部斜杠，请为单独Ingress设置注解：

```
nginx.ingress.kubernetes.io/preserve-trailing-slash: "true"
```

5.3.8 对接外部的认证服务

本示例中我们将使用https://httpbin.org/的认证接口，因为httpbin.org有一个认证接口为https://httpbin.org/basic-auth/{user}/{passwd}，这是一个认证地址，用户名是URL中的user，密码是passwd。请求示例如下：

```
$ curl -u admin:dean https://httpbin.org/basic-auth/admin/dean
{
  "authenticated": true,
  "user": "admin"
}
```

使用外部认证的Ingress：

```
apiVersion: networking.k8s.io/v1
kind: Ingress
metadata:
  name: ingress-echo-with-auth-basic-ext
  annotations:
    # 认证接口，设置了用户和密码
    nginx.ingress.kubernetes.io/auth-url: "https://httpbin.org/basic-auth/admin/dean"
spec:
  rules:
  - host: auth-basic-ext.deanit.cn
    http:
      paths:
      - path: /
        backend:
          service:
            # svc的名字
            name: echo
            port:
```

```
        # svc的端口
        number: 80
```

进行认证访问测试：

```
--- 使用错误的用户名或密码
$ curl -u user:nopass -H "Host: auth-basic-ext.deanit.cn" auth-basic-ext.deanit.cn
# 输出信息
<html>
<head><title>401 Authorization Required</title></head>
<body>
<center><h1>401 Authorization Required</h1></center>
<hr><center>openresty/1.15.8.1</center>
</body>
</html>


--- 使用正确的用户名和密码
$ curl -u admin:dean -H "Host: auth-basic-ext.deanit.cn" auth-basic-ext.deanit.cn
# 输出信息
Hostname: echo-597d89dcd9-4dp6f
```

由此可以看出，通过正确的用户名和密码进行认证可以成功访问服务。

5.3.9 配置默认页面

当Ingress资源中没有匹配到任何路由规则时，defaultBackend将作为默认的后端服务来处理这些请求。它允许用户定义一个默认的服务和端口，以便在请求无法匹配到任何具体规则时，仍然能够返回一个有效的响应，而不是错误或默认的404页面。Ingress配置示例如下：

```
apiVersion: networking.k8s.io/v1
kind: Ingress
metadata:
  name: deanit
  namespace: dean
spec:
  rules:
  - host: "www.deanit.cn"
    http:
      paths:
      - pathType: Prefix
        path: "/"       # 如果这个页面访问不到了，就会显示默认页面
        backend:
          service:
            name: deanit
            port:
              number: 80
  # 默认页面配置
  defaultBackend:
    service:
      # 默认页面服务的svc
```

```
      name: deanitrequest
      # 默认页面服务的端口
      port:
        number: 80
```

5.3.10　Nginx 如何获取客户端真实 IP

在Kubernetes中，当一个Pod接收到由另一个Pod转发的请求时，转发请求的Pod看到的客户端IP实际上是原始请求Pod的IP地址。这种情况可能导致服务无法获取客户端的真实IP地址，从而引发问题。为了解决这一问题，可以使用externalTrafficPolicy参数。

externalTrafficPolicy的优势在于，它允许Pod内的应用获取到真实的客户端IP地址。然而，其缺点是客户端只能使用Pod所在节点的IP进行访问，无法使用其他节点的IP或者通过vrrp等虚拟IP进行访问，这可能会给客户端带来一些不便。要实现这一点，可以将service.spec.externalTrafficPolicy的值设置为Local，这样请求只会被代理到本地端点，而不会被转发到其他节点，从而保留了最初的源IP地址。

Kubernetes会在Pod所在节点上针对nodePort下发DNAT规则，而在其他节点上针对nodePort下发DROP规则。以下是一个Service资源清单配置示例：

```
kind: Service
apiVersion: v1
metadata:
  name: deanit
  namespace: dean
spec:
  type: ClusterIP
  externalTrafficPolicy: Local
  ports:
    - port: 80
      targetPort: 80
  selector:
    app: deanit
```

5.3.11　重定向

在本示例中，我们将测试访问www.deanit.cn重定向到https://blog.deanit.cn。Ingress配置示例如下：

```
apiVersion: networking.k8s.io/v1
kind: Ingress
metadata:
  name: deanit-rewrite-target
  annotations:
    nginx.ingress.kubernetes.io/rewrite-target: https://blog.deanit.cn
spec:
  rules:
```

```
    - host: www.deanit.cn
      http:
        paths:
        - path: /
          backend:
            service:
              name: deanit
              port:
                number: 80
```

5.3.12 重写

Kubernetes中的Ingress重写功能主要用于根据特定的规则修改或替换请求的路径，以实现对不同后端服务的代理和访问。在下面的示例中，当请求路径以/backend/开头时，Ingress将对其进行重写，并将其转发到名为backend-service的后端服务的/newpath/{捕获的任意字符}路径上。Ingress配置示例如下：

```
apiVersion: networking.k8s.io/v1
kind: Ingress
metadata:
  name: my-ingress
  annotations:
    nginx.ingress.kubernetes.io/rewrite-target: /newpath/$1
spec:
  rules:
  - host: deanit.cn
    http:
      paths:
      - path: /backend/(.*)
        pathType: Prefix
        backend:
          service:
            name: backend-service
            port:
              number: 8080
```

5.3.13 多域名指向同一个后端服务

在生产环境中，一个门户网站可能需要多个域名绑定到服务上，这里就需要通过配置将多个域名写到Ingress的配置中，避免一个域名创建一个Ingress造成冗余。Ingress配置示例如下：

```
apiVersion: networking.k8s.io/v1
kind: Ingress
metadata:
  annotations:
    kubernetes.io/ingress.class: "nginx"
  name: demo3
  namespace: dean
```

```yaml
spec:
  rules:
  - host: demo3.deanit.cn
    http: &http_rules
      paths:
      - path: /
        backend:
          service:
            name: deanit
            port:
              number: 80
  # 绑定的其他域名
  - host: demo4.deanit.cn
    http: *http_rules
  - host: demo5.deanit.cn
    http: *http_rules
  - host: demo6.deanit.cn
    http: *http_rules
```

5.4 本章小结

本章介绍了Ingress-Nginx的基本概念和对应Pod的作用。首先，Ingress-Nginx作为Kubernetes中广泛使用的Ingress控制器之一，为外部流量进入集群内部服务提供了高效、灵活的路由机制。然后，详细介绍了如何在Kubernetes集群中安装和部署Ingress-Nginx，接下来，详细讲解了Ingress资源的定义与配置，包括如何设置HTTP路由规则、TLS证书以实现HTTPS访问，以及如何通过注解（Annotations）来定制Nginx的行为。通过这些内容的学习，读者将能够灵活配置Ingress-Nginx以满足不同场景下的需求。通过本章的学习和实践，读者可以对Ingress-Nginx有一个初步的了解，为后续章节的学习打下基础。

第 6 章

Kubernetes存储与持久化

在现代的云原生应用中,数据的持久化存储是一个关键问题。随着微服务架构和容器化技术的广泛应用,如何在Kubernetes环境中有效地管理存储资源成为开发者必须面对的挑战。本章将深入探讨Kubernetes存储的核心概念和实践方法,旨在帮助读者理解并掌握如何在Kubernetes集群中实现数据的持久化和高效管理。

本章首先介绍Kubernetes存储类的概念,这是理解和使用Kubernetes存储的基础。接着,详细阐述持久卷声明的重要性以及如何管理持久卷的生命周期。此外,通过动态申请持久卷的实验,读者将学习到如何在实际应用中利用存储类来动态分配存储资源,特别是通过NFS共享存储的方式。

本章还将对持久卷和持久卷声明进行详细的解析,包括它们的概述、创建流程、访问模式、回收策略以及卷状态等内容。这些知识点对于理解和使用Kubernetes存储至关重要。最后,本章还将介绍如何通过Deployment直接连接到NFS存储,这是一种常见的存储解决方案,适用于需要高可用性和持久性的应用。

6.1 Kubernetes存储类概述

在探索Kubernetes存储与持久化的奥秘时,我们不得不提及两个核心概念——存储类和持久卷。它们是容器化应用数据持久化的基础,为数据存储的安全性和可靠性提供了重要保障。

1. 存储类

存储类(Storage Class)是Kubernetes中用于描述持久卷(Persistent Volume)特性的API对象。它通常缩写为sc,是定义一组持久卷属性的关键工具。这些属性包括但不限于容量、存储类型、访问模式等,同时它还指定了负责提供这些持久卷的提供者(Provisioner)。当Pod请求存储资源时,

Kubernetes将依据存储类的定义动态地创建相应的持久卷，确保数据的高效存储与管理。

2. 持久卷

持久卷（Persistent Volume，PV）作为Kubernetes中的一个核心资源对象，专门用于在Pod中存储数据。Pod中的容器能够通过挂载（Mount）操作轻松访问持久卷中的数据。值得注意的是，持久卷的生命周期与Pod紧密相连。当Pod被删除时，若未采用持久化卷策略，持久卷中的数据可能会随之丢失。因此，合理配置和管理持久卷对于保障数据的安全性至关重要。

6.2 Kubernetes持久卷声明

在Kubernetes的存储管理领域，持久卷和持久卷声明（Persistent Volume Claim，PVC）是两个不可或缺的概念。它们共同构成了Kubernetes中数据持久化的核心机制，为容器化应用提供了稳定、可靠的数据存储解决方案。

1. 持久卷

持久卷是Kubernetes集群中对网络存储的一种抽象表示。它如同一个虚拟的存储空间，可以被挂载到一个或多个Pod上，作为这些Pod的持久化存储介质。与Pod不同的是，PV的生命周期独立于Pod存在。这意味着，即使Pod被销毁，持久卷中的数据依然会安然无恙地保留下来，确保数据的持久性。

2. 持久卷声明

持久卷声明则是用户对于存储资源的一种请求方式。它与Pod有着异曲同工之妙。Pod消耗的是节点上的计算资源，如CPU和内存，通过spec.containers[].resources.limits进行定义；而PVC则消耗的是PV提供的存储资源。用户可以根据自己的需求，通过PVC请求特定大小和访问模式的存储资源，以满足应用的存储需求。这种灵活性使得Kubernetes能够更好地适应不同应用的存储需求，实现资源的高效利用。

6.3 持久卷的生命周期

在Kubernetes的存储管理领域，持久卷和持久卷声明的生命周期是理解其工作机制的关键。它们共同构成了数据持久化的核心流程，确保了应用数据的可靠性和持久性。

1. 持久卷的生命周期

持久卷（PV）的生命周期通常涵盖以下几个关键阶段。

01 创建：管理员通过API服务器精心创建PV对象，为其设定独特的属性和特性。

02 绑定：用户通过PVC向Kubernetes发出存储资源请求，系统根据PVC的具体要求与现有PV的属性进行智能匹配。一旦找到合适的PV，Kubernetes便将其与PVC紧密绑定。

03 使用：Pod通过与其关联的PVC，顺畅地访问并使用PV提供的存储资源，确保数据的安全存储与高效读写。

04 释放：当Pod被删除且相应的PVC也随之消失时，PV进入"释放"状态，静待下一次与新的PVC建立连接的机会。

05 删除：管理员可通过API服务器主动删除PV对象，从而回收其占用的资源。

2. 持久卷声明的生命周期

持久卷声明（PVC）的生命周期包括以下步骤。

01 创建：用户通过API服务器创建PVC对象，明确表达对存储资源的需求。

02 绑定：Kubernetes系统自动根据PVC的详细要求，在现有的PV中寻找与之匹配的对象。一旦匹配成功，立即建立PVC与PV之间的绑定关系。

03 使用：Pod通过与其绑定的PVC，无缝地访问并利用PV提供的存储资源，实现数据的持久化存储。

04 删除：当PVC不再满足用户需求或任务完成时，用户可以主动通过API服务器删除PVC对象。此时，与该PVC绑定的PV将被释放，重新进入可用资源池，等待下一次被绑定的机会。

6.4 动态申请持久卷实验

本节将通过实验深入探讨如何利用NFS（Network File System）共享存储，创建存储类，以及如何使用这些存储类来动态申请资源，从而为应用提供稳定、高效的数据存储解决方案。

6.4.1 NFS 共享存储搭建

NFS是一种分布式文件系统协议，它允许计算机客户端通过网络访问服务器上的文件，就像访问本地文件系统一样。NFS最初由Sun Microsystems开发，现在已经成为UNIX和Linux系统中广泛使用的标准之一。

在本示例中，我们将规划使用k8s-master节点（IP：192.168.1.10）作为NFS服务端，以供集群使用持久化存储。在生产环境中，一般使用单独的存储服务器来作为后端存储。

具体操作步骤如下：

01 在k8s-master节点安装NFS服务软件包：

```
[root@k8s-master ~]# yum install -y nfs-utils rpcbind
```

该命令从默认的软件源中安装以下软件包。

- nfs-utils：是一个工具包，用于配置和管理NFS服务器和客户端。NFS是一种允许不同计算机通过网络共享文件和目录的协议。通过nfs-utils，用户可以轻松地在网络上的计算机之间共享文件，无须手动复制文件到每台计算机。
- rpcbind：通过管理RPC的注册信息和端口映射来支持远程过程调用和通信。

02 在k8s-master节点创建存储目录并赋值权限：

```
--- 创建NFS存储目录
[root@k8s-master ~]# mkdir /nfs

--- 赋予权限
[root@k8s-master ~]# chmod 777 /nfs

--- 赋予所属组和所属用户
[root@k8s-master ~]# chown nfsnobody.nfsnobody /nfs
```

上述命令用于在Linux系统中创建并更改/nfs目录的所有者和所属组为nfsnobody用户和组以及权限。这通常是在配置NFS服务器时进行的操作，特别是当你希望所有对/nfs目录的访问都通过NFS服务器的用户身份管理时。

03 在k8s-master节点修改NFS服务配置文件：

```
[root@k8s-master ~]# vim /etc/exports
/nfs *(rw,no_root_squash,no_all_squash,sync)
```

该配置文件参数表示/nfs目录将被导出为NFS共享，并且允许所有客户端以读写方式访问。当设置no_root_squash时，NFS客户端上的root用户将具有NFS服务器上共享目录的root用户权限。这允许root用户进行需要超级用户权限的操作。启用no_all_squash选项时，NFS服务器不会将所有远程访问的普通用户及其所属组映射为匿名用户（通常是nfsnobody用户或组）。相反，它将保留共享文件的原始UID（User Identifier，用户标识符）和GID（Group Identifier，组标识符）。这意味着，如果客户端用户具有对服务器上文件的特定权限，他们将在访问这些文件时保持这些权限，而不是被降级为匿名用户。sync选项表示写入操作将被同步到磁盘。

04 在k8s-master节点设置NFS服务开机自启并启动：

```
--- 设置开机自启和启动，严格按照启动顺序(rpcbind, nfs)
systemctl enable rpcbind --now
systemctl enable nfs --now
```

这两个命令用于在Linux系统中禁用和停止firewalld服务。

- systemctl enable rpcbind --now：这个命令的作用是将rpcbind服务设置为开机自启，--now选项是一个快捷方式，用于在启用服务的同时立即开始运行。换句话说，它等同于先运行systemctl enable rpcbind（启用服务以便在系统启动时自动运行），紧接着运行systemctl start rpcbind（立即启动服务）。
- systemctl enable nfs --now：这个命令的作用是将NFS服务设置为开机自启，并使用--now将NFS服务立即启动。

05 在k8s-master节点查看NFS服务状态：

```
systemctl status rpcbind nfs
```

该命令用于查看rpcbind和nfs服务，以确保服务启动正常。

06 在k8s-node1、k8s-node2、k8s-node3节点上安装NFS服务软件包：

```
yum install -y nfs-utils rpcbind
```

07 在k8s-node1、k8s-node2、k8s-node3节点设置NFS服务开机自启并启动：

```
--- 设置开机自启和启动，严格按照启动顺序(rpcbind, nfs)
systemctl enable rpcbind --now
systemctl enable nfs --now
```

08 在k8s-node1、k8s-node2、k8s-node3节点查看NFS服务状态：

```
systemctl status rpcbind nfs
```

09 在k8s-node1节点测试连接NFS服务端，通过结果可以看到NFS服务端的存储目录：

```
[root@k8s-node1 ~]# showmount -e 192.168.1.10
Export list for 192.168.1.10:
/nfs *
```

该命令用于显示NFS服务器上（在本例中是IP地址为192.168.1.10的服务器）导出的所有共享目录的信息。这个命令通常在NFS客户端上执行，用于检查NFS服务器上配置了哪些共享资源，以及这些资源的导出选项。

至此，NFS服务搭建完成。

6.4.2 nfs-client-provisioner 存储类搭建

nfs-client-provisioner是Kubernetes生态系统中的一个关键组件，它负责在NFS共享存储上动态地创建和管理持久卷。在Kubernetes集群中，当Pod需要持久化存储时，它会通过持久卷声明向系统请求所需的存储资源。而nfs-client-provisioner则扮演着"中介"的角色，根据PVC的请求，自动在NFS共享存储上创建相应的PV，并将其绑定到PVC上。这样，Pod就可以通过PVC访问并使用PV提供的存储资源了。本小节将详细介绍如何搭建nfs-client-provisioner存储类，并解决可能出现的问题。

接下来，我们将介绍nfs-client-provisioner存储类的搭建步骤。

01 在k8s-master节点创建资源清单：

```
[root@k8s-master ~]# vim nfs-client-provisioner.yaml
---
apiVersion: v1
kind: Namespace
metadata:
  name: nfs-client-provisioner
  labels:
    name: nfs-client-provisioner
---
apiVersion: storage.k8s.io/v1
kind: StorageClass
metadata:
  name: dean-nfs            # sc的名字
# 这个要和nfs-client-provisioner的env环境变量中的PROVISIONER_NAME的value值对应。Pod使用不
同资源可以用这个来区分
provisioner: dean-nfs
reclaimPolicy: Retain       # 回收策略为Retain（手动释放），删除pv将不会删除服务器上的文件
---
apiVersion: v1
kind: ServiceAccount
metadata:
  name: nfs-client-provisioner         # sa名字
  namespace: nfs-client-provisioner    # 命名空间
---
# 创建集群规则
kind: ClusterRole
apiVersion: rbac.authorization.k8s.io/v1
metadata:
  name: nfs-client-provisioner-runner
rules:
  - apiGroups: [""]
    resources: ["persistentvolumes"]
    verbs: ["get", "list", "watch", "create", "delete"]
  - apiGroups: [""]
    resources: ["persistentvolumeclaims"]
    verbs: ["get", "list", "watch", "update"]
  - apiGroups: ["storage.k8s.io"]
    resources: ["storageclasses"]
    verbs: ["get", "list", "watch"]
  - apiGroups: [""]
    resources: ["events"]
    verbs: ["create", "update", "patch"]
---
# 将服务认证用户与集群规则进行绑定
kind: ClusterRoleBinding
apiVersion: rbac.authorization.k8s.io/v1
```

```yaml
metadata:
  name: run-nfs-client-provisioner
subjects:
  - kind: ServiceAccount                            # 类型为sa
    name: nfs-client-provisioner                    # sa的名字一致
    namespace: nfs-client-provisioner               # 和nfs provisioner安装的namespace一致
roleRef:
  kind: ClusterRole
  name: nfs-client-provisioner-runner
  apiGroup: rbac.authorization.k8s.io
---
kind: Role
apiVersion: rbac.authorization.k8s.io/v1
metadata:
  name: leader-locking-nfs-client-provisioner
  namespace: nfs-client-provisioner                 # 和nfs provisioner安装的namespace一致
rules:
  - apiGroups: [""]
    resources: ["endpoints"]
    verbs: ["get", "list", "watch", "create", "update", "patch"]
---
kind: RoleBinding
apiVersion: rbac.authorization.k8s.io/v1
metadata:
  name: leader-locking-nfs-client-provisioner
  namespace: nfs-client-provisioner                 # 和nfs provisioner安装的namespace一致
subjects:
  - kind: ServiceAccount                            # 类型为sa
    name: nfs-client-provisioner                    # sa的名字一致
    namespace: nfs-client-provisioner               # 和nfs provisioner安装的namespace一致
roleRef:
  kind: Role
  name: leader-locking-nfs-client-provisioner
  apiGroup: rbac.authorization.k8s.io
---
apiVersion: apps/v1
kind: Deployment
metadata:
  name: nfs-client-provisioner
  labels:
    app: nfs-client-provisioner
  namespace: nfs-client-provisioner                 # 部署在指定ns下
spec:
  replicas: 1                                       # 副本数,建议为奇数[1,3,5,7,9]
  strategy:
    type: Recreate                                  # 使用重建的升级策略
  selector:
    matchLabels:
      app: nfs-client-provisioner
  template:
```

```yaml
  metadata:
    labels:
      app: nfs-client-provisioner
  spec:
    serviceAccountName: nfs-client-provisioner     # sa名字
    containers:
      - name: nfs-client-provisioner               # 容器名字
        image: quay.io/external_storage/nfs-client-provisioner:latest
        volumeMounts:
          - name: nfs-client-root
            mountPath: /persistentvolumes          # 指定容器内挂载的目录
        env:
          - name: PROVISIONER_NAME                 # 容器内的变量用于指定提供存储的名称
            # nfs-client-provisioner的名称，以后设置的storage class要和这个保持一致
            value: dean-nfs
          - name: NFS_SERVER        # 容器内的变量指定nfs服务器对应的目录
            value: 192.168.1.10     # NFS服务器的地址
          - name: NFS_PATH          # 容器内的变量指定nfs服务器对应的目录
            value: /nfs   # NFS服务的挂载目录,如果采用这个nfs动态申请PV,所创建的文件在这个目录下
        volumes:
          - name: nfs-client-root       # 赋值卷名字
            nfs:
              server: 192.168.1.10      # NFS服务器的地址
              path: /nfs      # NFS服务的挂载目录,通常要给权限,生产时可根据实际情况而定
```

02 应用资源：

```
[root@k8s-master ~]        # kubectl apply -f nfs-client-provisioner.yaml
namespace/nfs-client-provisioner created
storageclass.storage.k8s.io/dean-nfs created
serviceaccount/nfs-client-provisioner created
clusterrole.rbac.authorization.k8s.io/nfs-client-provisioner-runner created
clusterrolebinding.rbac.authorization.k8s.io/run-nfs-client-provisioner created
role.rbac.authorization.k8s.io/leader-locking-nfs-client-provisioner created
rolebinding.rbac.authorization.k8s.io/leader-locking-nfs-client-provisioner created
deployment.apps/nfs-client-provisioner created
```

该命令使用Kubectl工具将nfs-client-provisioner.yaml文件中定义的资源应用到Kubernetes集群中。使用-f选项指定要应用的文件名。根据输出信息，可以看出已经成功创建了以下资源：

```
namespace/nfs-client-provisioner created
storageclass.storage.k8s.io/dean-nfs created
serviceaccount/nfs-client-provisioner created
clusterrole.rbac.authorization.k8s.io/nfs-client-provisioner-runner created
clusterrolebinding.rbac.authorization.k8s.io/run-nfs-client-provisioner created
role.rbac.authorization.k8s.io/leader-locking-nfs-client-provisioner created
rolebinding.rbac.authorization.k8s.io/leader-locking-nfs-client-provisioner created
```

```
deployment.apps/nfs-client-provisioner created
```

03 查看Pod资源状态：

```
[root@k8s-master ~]# kubectl get pods -n nfs-client-provisioner
NAME                                       READY   STATUS    RESTARTS   AGE
nfs-client-provisioner-7cd5787dd7-x28hp    1/1     Running   0          27s
```

04 查看存储类资源信息：

```
[root@k8s-master ~]# kubectl get sc
NAME       PROVISIONER   RECLAIMPOLICY   VOLUMEBINDINGMODE   ALLOWVOLUMEEXPANSION   AGE
dean-nfs   dean-nfs      Retain          Immediate           false                  31m
```

参数解析：

- NAME：存储类名称，在上面的例子中，服务的名字是dean-nfs。
- PROVISIONER：动态创建持久卷的存储类供应者。
- RECLAIMPOLICY：定义了当持久卷的引用被删除时的处理策略。具体选项包括：
    - Retain：手动回收。
    - Recycle：（已弃用，不推荐使用）
    - Delete：删除与该PV关联的底层存储。
- VOLUMEBINDINGMODE：定义了PV和PVC之间的绑定模式。对于动态存储供应尤为重要。
- ALLOWVOLUMEEXPANSION：这是一个布尔值，指示是否允许扩展已存在的PV的大小。如果设置为true，并且底层存储支持扩展，则可以通过编辑PVC来请求更大的存储大小。
- AGE：表示资源在Kubernetes集群中的存在时间。

05 申请PVC会出现以下报错：

```
waiting for a volume to be created, either by external provisioner "dean-nfs" or manually created by system administrator
```

06 修改APIServer配置来禁用RemoveSelfLink功能，从而继续保留selfLink字段：

```
[root@k8s-master ~]# vim /etc/kubernetes/manifests/kube-apiserver.yaml
# 在以下层级下添加
spec:
 containers:
 - command:
# 添加如下内容：--feature-gates=RemoveSelfLink=false，这将关闭RemoveSelfLink特性门，从而开启selfLink属性
 - --feature-gates=RemoveSelfLink=false
```

运维笔记：修改保存后，APIServer将会自动重启，等待5分钟左右再次执行kubectl get node来确认集群是否恢复正常。

6.4.3 服务使用存储类动态申请资源实验

本实验将演示如何在k8s-master节点上创建和使用一个名为mynginx的服务，并使用StatefulSet和PVC来动态申请和管理存储资源。我们将通过定义资源清单文件和应用这些资源，展示如何实现这一过程。

01 在k8s-master节点创建资源清单。

首先，创建一个包含Service和StatefulSet定义的YAML文件：

```
[root@k8s-master ~]# vim mynginx.yaml
apiVersion: v1
kind: Service
metadata:
  name: mynginx
  namespace: dean
  labels:
    app: mynginx
spec:
  ports:
  - port: 80
    name: web
  clusterIP: None
  selector:
    app: mynginx
---
apiVersion: apps/v1
kind: StatefulSet
metadata:
  name: mynginx
  namespace: dean
spec:
  selector:
    matchLabels:
      app: mynginx
  serviceName: "mynginx"
  replicas: 3
  template:
    metadata:
      labels:
        app: mynginx
    spec:
      containers:
      - name: mynginx
        image: nginx:1.21
        ports:
        - containerPort: 80
          name: web
        # 卷挂载信息
```

```
          volumeMounts:
          # 使用www名称的卷模板名
          - name: www
            # 挂载到Pod中的路径
            mountPath: /usr/share/nginx/html
  # volumeClaimTemplates是在StatefulSet资源的定义中用于创建 PersistentVolumeClaims(PVCs)
的模板
  volumeClaimTemplates:
  - metadata:
      # PVC的名字是www，在StatefulSet中，每个Pod的PVC名称都会基于这个模板名称和Pod的序号进行
扩展
      # 例如，如果StatefulSet的名字是mynginx，并且它有3个Pod，那么PVC的名称是www-mynginx-0、
www-mynginx-1和www-mynginx-2
      name: www
    spec:
      # 定义了存储的访问模式。在这个例子中，访问模式是ReadWriteOnce，这意味着PVC可以被单个节点
以读写方式挂载
      accessModes: [ "ReadWriteOnce" ]
      # 指定了PVC使用的存储类 (StorageClass)
      storageClassName: "dean-nfs"
      # 定义了请求的存储大小
      resources:
        requests:
          storage: 1Gi
```

02 应用资源。

使用Kubectl工具将上述YAML文件中定义的资源应用到Kubernetes集群中：

```
[root@k8s-master ~]# kubectl apply -f mynginx.yaml
service/mynginx created
statefulset.apps/web created
```

该命令成功创建了以下资源：

- service/mynginx created
- statefulset.apps/web created

03 查看Pod和Service信息。

使用kubectl get命令查看Pod和Service的状态：

```
--- Pod信息
[root@k8s-master ~]# kubectl get pods -n dean | grep mynginx
mynginx-0              1/1      Running      0           6m39s
mynginx-1              1/1      Running      0           41s
mynginx-2              1/1      Running      0           35s

--- Service信息
[root@k8s-master ~]# kubectl get svc -n dean | grep mynginx
mynginx         ClusterIP    None           <none>         80/TCP      6m39s
```

04 查看mynginx服务的PV和PVC资源信息

为了验证PV和PVC的创建情况，可以执行如图6-1所示的命令。

```
[root@k8s-master ~]# kubectl get pv
NAME                                       CAPACITY   ACCESS MODES   RECLAIM POLICY   STATUS   CLAIM                  STORAGECLASS   REASON   AGE
pvc-50344981-ec0b-47c8-85ee-4ecd8f68a365   1Gi        RWO            Retain           Bound    dean/www-mynginx-0     dean-nfs                112s
pvc-54130a68-80fc-4271-adc9-129785cabbc3   1Gi        RWO            Retain           Bound    dean/www-mynginx-2     dean-nfs                78s
pvc-9e060049-a793-41a7-be8a-eff5ede6ef4b   1Gi        RWO            Retain           Bound    dean/www-mynginx-1     dean-nfs                85s
[root@k8s-master ~]#
[root@k8s-master ~]#
[root@k8s-master ~]# kubectl get pvc -A
NAMESPACE   NAME              STATUS   VOLUME                                     CAPACITY   ACCESS MODES   STORAGECLASS   AGE
dean        www-mynginx-0     Bound    pvc-50344981-ec0b-47c8-85ee-4ecd8f68a365   1Gi        RWO            dean-nfs       7m28s
dean        www-mynginx-1     Bound    pvc-9e060049-a793-41a7-be8a-eff5ede6ef4b   1Gi        RWO            dean-nfs       90s
dean        www-mynginx-2     Bound    pvc-54130a68-80fc-4271-adc9-129785cabbc3   1Gi        RWO            dean-nfs       84s
```

图 6-1　查看 mynginx 服务的 PV 和 PVC 信息的效果

在k8s-master节点查看NFS服务端的存储目录，可以看到服务创建的持久化目录，如图6-2所示。

```
[root@k8s-master ~]# ll /nfs
总用量 0
drwxrwxrwx 2 root root 6 6月  21 17:17 dean-www-mynginx-0-pvc-50344981-ec0b-47c8-85ee-4ecd8f68a365
drwxrwxrwx 2 root root 6 6月  21 17:18 dean-www-mynginx-1-pvc-9e060049-a793-41a7-be8a-eff5ede6ef4b
drwxrwxrwx 2 root root 6 6月  21 17:18 dean-www-mynginx-2-pvc-54130a68-80fc-4271-adc9-129785cabbc3
```

图 6-2　查看 NFS 服务端存储目录

至此，服务使用存储类动态申请资源实验完成。通过创建Service和StatefulSet资源，并配置PVC和StorageClass，我们实现了对持久化数据的高效管理和自动扩展。

6.5　PV/PVC详解

本节将详细介绍Kubernetes中的PV和PVC的概念、创建流程、访问模式、回收策略以及卷状态。

6.5.1　PV/PVC 概述

PV是由管理员设置的存储，它是群集的一部分。就像节点是集群中的资源一样，PV也是集群中的资源。PV是Volume之类的卷插件，但具有独立于使用PV的Pod的生命周期。此API对象包含存储实现的细节，即NFS、iSCSI或特定于云供应商的存储系统。

PVC是用户存储的请求。它与Pod相似。Pod消耗节点资源，PVC消耗PV资源。Pod可以请求特定级别的资源（CPU和内存）。声明可以请求特定的大小和访问模式（例如，可以以读/写一次或只读多次模式挂载）。

6.5.2　PVC 的创建流程

PVC的创建流程在Kubernetes中通常涉及几个关键步骤。以下是PVC创建流程的图解，如图6-3所示。

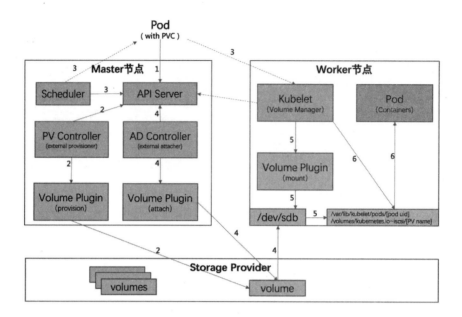

图 6-3 PVC 创建流程图解

6.5.3 PV 访问模式

PersistentVolume能够以资源提供者支持的任何方式挂载到主机上。如表6-1所示，供应商具有不同的功能，每个PV的访问模式都将被设置为该卷支持的特定模式。例如，NFS可以支持多个读/写客户端，但特定的NFS PV只能以只读方式导出到服务器上。每个PV都有一套自己的用来描述特定功能的访问模式。

表 6-1 PV 访问模式

访问模式	定　　义	缩　　写
ReadWriteOnce	该卷可以被单个节点以读/写模式挂载	RWO
ReadOnlyMany	该卷可以被多个节点以只读模式挂载	ROX
ReadWriteMany	该卷可以被多个节点以读/写模式挂载	RWX

6.5.4 PV 回收策略

PV回收策略主要涉及如何管理和回收集群中的资源，以确保资源的有效利用和避免浪费。

PV回收策略通过persistentVolumeReclaimPolicy字段来定义，该字段有以下可选值。

- Retain：保留PV，不进行自动回收。当PV使用完成后，需要管理员手动进行清理和释放。
- Delete：删除PV。当PV不再被使用时，Kubernetes会自动删除并释放它。请注意，此操作会删除PV对象及其对应的存储资源，所以需确保在使用此策略前已经备份了重要数据。

- Recycle（已废弃）：回收PV。在Kubernetes v1.14版本之前，该策略会尝试清空PV中的数据，但不保证数据安全。由于存在安全风险，此策略在v1.14版本后已被废弃，不再推荐使用。

6.5.5 PV/PVC 卷状态

在Kubernetes中，PV和PVC是用于管理持久化存储资源的核心概念。PV是集群中的一块网络存储，而PVC则是对PV的请求。

PV在Kubernetes中可能处于以下几种状态：

- Available（可用）。
  - 描述：PV已经被Kubernetes集群管理员创建，并且还没有被绑定到任何PVC上，可以被任何PVC请求使用。
  - 触发条件：管理员创建PV后，其默认状态即为Available。
- Bound（已绑定）。
  - 描述：PV已经被绑定到一个PVC上，可以被挂载到一个Pod中使用。
  - 触发条件：当PVC被创建后，Kubernetes会尝试将其绑定到一个可用的PV上。如果有可用的PV，则PVC会被绑定到该PV上，PV的状态会变为Bound。
- Released（已释放）。
  - 描述：PVC与PV之间的绑定关系已经被删除，但是PV上的数据还没有被清除，这时PV处于Released状态，可以被重新绑定到另一个PVC上使用。
  - 触发条件：当PVC被删除，但PV上的数据被保留时，PV的状态会变为Released。
- Failed（失败）。
  - 描述：PV与底层存储后端的连接出现问题，或者存储后端出现了错误，导致PV无法使用，这时PV处于Failed状态。
  - 触发条件：PV与底层存储系统之间的通信问题或存储系统本身的问题可能导致PV进入Failed状态。

PVC的状态通常与PV的绑定过程相关：

- Pending（等待中）。
  - 描述：PVC已经被创建，但尚未找到可用的PV进行绑定。
  - 触发条件：当PVC被创建后，Kubernetes会尝试将其绑定到一个满足其要求的PV上。如果没有合适的PV可用，PVC将处于Pending状态。
- Bound（已绑定）。
  - 描述：PVC已经成功绑定到一个PV上，可供Pod使用。
  - 触发条件：当PVC成功绑定到一个PV时，其状态会变为Bound。

- Lost（丢失）。
  - ◆ 描述：PVC与PV之间的绑定关系已经丢失，这通常发生在PV被意外删除或集群状态不一致时。
  - ◆ 触发条件：PV被删除或集群状态不一致可能导致PVC进入Lost状态。

我们可以对PV/PVC进行以下管理和操作。

- 创建PV：管理员可以创建PV，并指定其属性，如存储类、容量、访问模式等。创建后，PV处于Available状态。
- 创建PVC：用户可以创建PVC，并指定需要的存储容量、存储类和访问模式等属性。当PVC被创建后，Kubernetes会尝试将其绑定到一个可用的PV上。
- 手动绑定：管理员也可以手动将一个Available状态的PV绑定到一个PVC上。
- 解绑定和删除：当PVC与PV之间的关联关系不再需要时，可以将它们解绑定。如果PV的状态为Released，则可以直接删除它；如果PV的状态为Bound，则需要先解绑定它，再删除它。

6.6 Deployment直连NFS存储

在Kubernetes中，我们可以让Deployment类型的服务直接连接NFS存储，从而避免创建PV/PVC资源。

运维笔记：这种直接连接NFS的方法应仅作为特殊情况下的解决方案，在实际生产环境中，建议使用存储类来动态申请和管理存储资源。

在NFS存储服务端创建持久化目录，按照6.4.1节的规划，在k8s-master节点上进行操作：

```
[root@k8s-master ~]# mkdir /nfs/deploy-nfs-pressure
```

资源清单示例：

```
apiVersion: apps/v1
kind: Deployment
metadata:
  namespace: dean
  name: deploy-nfs-pressure
spec:
  replicas: 3
  revisionHistoryLimit: 10
  selector:
    matchLabels:
      app: nginx-nfs
  template:
    metadata:
      labels:
```

```yaml
      app: nginx-nfs
  spec:
    containers:
      - name: nginx
        imagePullPolicy: IfNotPresent
        image: nginx:1.21
        ports:
          - containerPort: 80
            name: web-nginx
        volumeMounts:
          - mountPath: /usr/share/nginx/html
            name: web-path
    volumes:
      - name: web-path
        # 这里选择为nfs存储模式
        nfs:
          # nfs服务端的目录
          path: /nfs/deploy-nfs-pressure
          # nfs服务端的IP
          server: 192.168.1.10
```

通过该资源清单创建的服务，多个Pod副本将使用一个存储目录，我们将前端文件存放在该目录下，即可实现多Pod副本前端文件更新。

6.7 本章小结

本章通过深入探讨在Kubernetes环境中使用nfs-client-provisioner作为动态存储类服务的实践，为读者构建了一个全面理解和操作持久化存储的框架。通过本章的学习，读者不仅能够深刻理解Kubernetes中存储卷、持久化卷和持久化卷声明的基本概念与作用，还能够掌握如何配置和使用nfs-client-provisioner来实现动态的存储供应。这将为构建高可用性、可扩展的Kubernetes应用奠定坚实的基础。无论是初学者还是经验丰富的Kubernetes运维人员，都能从中获得宝贵的知识和经验。

第 7 章

ConfigMap配置和Secret密钥管理

在Kubernetes中，ConfigMap和Secret是用于存储配置信息的两种重要资源类型。这些资源提供了将配置信息与应用程序代码分离的机制，使得应用程序的配置更加灵活和可管理。本章将详细介绍ConfigMap和Secret的基本概念、用途、创建方式以及使用场景。

7.1 ConfigMap：非敏感配置信息的集中管理

本节将详细介绍ConfigMap的基本概念、用途、创建方式以及使用场景。

7.1.1 ConfigMap 概述

ConfigMap是Kubernetes中的一种API对象，旨在将配置数据注入容器中。简称为cm的ConfigMap，使得应用程序的配置信息（例如数据库连接详情、环境变量等）能够在Kubernetes集群中得到存储，并在Pod启动时自动注入相应的容器中。这种机制通过键—值对的形式来组织数据，并且支持从文件或目录直接创建ConfigMap。

ConfigMap的主要用途如下：

- 作为环境变量的来源，供Pod中的容器使用。
- 提供命令行参数，允许在容器启动时指定特定的配置选项。
- 将整个配置文件存储到Kubernetes集群中，并在Pod启动时将其挂载到容器内。

ConfigMap适用于以下场景：

- 若应用程序的配置信息需要频繁更新，则可将其存储到ConfigMap中，并通过滚动更新Pod的方式来应用新的配置。
- 当多个Pod需要共享相同的配置信息时，利用ConfigMap可以避免在每个Pod中重复进行配置。

ConfigMap热更新解析：

- 使用ConfigMap挂载的环境变量不会同步更新。
- 使用ConfigMap挂载的卷中的数据需要一段时间（实测大约10秒）才能同步更新。
- 以subPath形式挂载的ConfigMap不会同步更新。

7.1.2 使用目录方式创建 ConfigMap

使用目录方式创建ConfigMap的步骤如下：

01 在k8s-master节点的root目录下创建configmap-dir目录并进入该目录：

```
[root@k8s-master ~]# mkdir /root/configmap-dir && cd /root/configmap-dir
```

02 在k8s-master节点创建名为game.properties的配置文件：

```
[root@k8s-master ~]# vim game.properties
# 内容如下：
enemies=aliens
lives=3
enemies.cheat=true
enemies.cheat.level=noGoodRotten secret.code.passphrase=UUDDLRLRBABAS
secret.code.allowed=true
secret.code.lives=30
```

03 在k8s-master节点创建名为ui.properties的配置文件：

```
[root@k8s-master ~]# vim ui.properties
# 内容如下：
color.good=purple
color.bad=yellow
allow.textmode=true
how.nice.to.look=fairlyNice
```

04 基于目录创建ConfigMap：

```
[root@k8s-master ~]# kubectl create configmap game-config --from-file=/root/configmap-dir/ -n dean
configmap/game-config created
```

该命令用于在Kubernetes集群中名为dean的命名空间中创建一个名为game-config的ConfigMap，并且这个ConfigMap的内容将从本地目录/root/configmap-dir/中的文件自动填充。--from-file指定在目录下的所有文件都将用于在ConfigMap中创建键一值对，其中键的名称就是文件名，而值则为文件的内容。

05 查看创建的ConfigMap：

```
[root@k8s-master ~]# kubectl get cm -n dean
```

```
NAME         DATA   AGE
game-config  2      20s
```

06 查看game-config详细配置：

```
[root@k8s-master ~]# kubectl get cm -n dean game-config -o yaml
apiVersion: v1
data:
  game.properties: "enemies=aliens\nlives=3\nenemies.cheat=true\
nenemies.cheat.level=noGoodRotten
    secret.code.passphrase=UUDDLRLRBABAS \nsecret.code.allowed=true
\nsecret.code.lives=30\n"
  ui.properties: |
    color.good=purple
    color.bad=yellow
    allow.textmode=true
    how.nice.to.look=fairlyNice
kind: ConfigMap
...
```

该命令用于查看Kubernetes集群中dean命名空间下名为game-config的ConfigMap配置信息并以YAML格式进行展示。

07 使用describe命令也可以看到配置信息：

```
[root@k8s-master ~]# kubectl describe cm -n dean game-config
Name:         game-config
Namespace:    dean
Labels:       <none>
Annotations:  <none>

Data
====
ui.properties:
----
color.good=purple
color.bad=yellow
allow.textmode=true
how.nice.to.look=fairlyNice

game.properties:
----
enemies=aliens
lives=3
enemies.cheat=true
enemies.cheat.level=noGoodRotten secret.code.passphrase=UUDDLRLRBABAS
secret.code.allowed=true
secret.code.lives=30

Events:  <none>
```

7.1.3 使用文件方式创建 ConfigMap

在k8s-master节点创建Nginx配置文件：

```
[root@k8s-master ~]# vim /root/nginx.conf
server {
  server_name www.deanit.cn;
  listen 80;
  root /home/nginx/www/
}
```

根据文件创建ConfigMap，定义一个键为www，其值为nginx.conf文件中的内容：

```
[root@k8s-master ~]# kubectl create configmap www-nginx --from-file=www=/root/nginx.conf -n dean
```

查看ConfigMap的详细信息：

```
[root@k8s-master ~]# kubectl describe configmap www-nginx -n dean
Name:          www-nginx
Namespace:     dean
Labels:        <none>
Annotations:   <none>

Data
====
www:
----
server {
  server_name www.deanit.cn;
  listen 80;
  root /home/nginx/www/
}
```

7.1.4 使用字面值方式创建 ConfigMap

基于字面值创建ConfigMap，利用--from-literal参数传递配置信息，该参数可以使用多次：

```
[root@k8s-master ~]# kubectl create configmap dean-info --from-literal=dean.city=beijing --from-literal=dean.name=dean -n dean
configmap/dean-info created
```

参数解析：

- dean-info：ConfigMap的名称。
- --from-literal=dean.city：指定了一个键名，名为dean.city。
- beijing：值。
- --from-literal=dean.name：又指定了一个键名，名为dean.name。
- dean：值。

使用describe命令也可以看到配置信息：

```
[root@k8s-master ~]# kubectl describe cm dean-info -n dean

Name:         dean-info
Namespace:    dean
Labels:       <none>
Annotations:  <none>

Data
====

dean.city:
----
beijing

dean.name:
----
dean
Events:  <none>
```

7.1.5 设置 ConfigMap 不允许更改

使用immutable字段设置当前ConfigMap是否允许被更改，下面是一个ConfigMap示例：

```
apiVersion: v1              # 版本
kind: ConfigMap             # 类型
immutable: true             # 若设置为true，则ConfigMap不允许被更改。默认是false
metadata:
  name: dean-cm             # 名字
  namespace: dean           # 命名空间
data:
  log_level: INFO
```

7.1.6 通过 envfrom 方式指定 ConfigMap

在k8s-master节点创建名为MySQL-Config的ConfigMap，其中定义了两个KEY：

```
[root@k8s-master ~] # vim mysql_config.yaml
apiVersion: v1
kind: ConfigMap
metadata:
  name: mysql-config
  namespace: dean
data:
  master.cnf: |
    [mysqld]
    datadir=/var/lib/mysql
```

```
    socket=/var/lib/mysql/mysql.sock
    symbolic-links=0
    server-id=1
  slave.cnf: |
    [mysqld]
    datadir=/var/lib/mysql
    socket=/var/lib/mysql/mysql.sock
    symbolic-links=0
    server-id=2
```

创建Deployment，通过envfrom方式指定ConfigMap，这种方式会引用ConfigMap中所有的KEY为系统变量，YAML资源清单如下：

```
[root@k8s-master ~]# vim envfrom_demo.yaml
apiVersion: apps/v1
kind: Deployment
metadata:
  name: envfrom-demo
  namespace: dean
spec:
  replicas: 1
  selector:
    matchLabels:
      name: envfrom-demo
  template:
    metadata:
      labels:
        name: envfrom-demo
    spec:
      containers:
      - name: envfrom-demo
        image: busybox:1.28.0
        imagePullPolicy: IfNotPresent
        command: ["/bin/sh", "-c", "sleep 360000"]
        envFrom:
        - configMapRef:
            name: mysql-config     # 指定ConfigMap名称
```

应用资源：

```
[root@k8s-master ~]# kubectl apply -f mysql_config.yaml
configmap/mysql-config created
[root@k8s-master ~]# kubectl apply -f envfrom_demo.yaml
deployment.apps/envfrom-demo created
```

该命令使用Kubectl工具将mysql_config.yaml、envfrom_demo.yaml文件中定义的资源应用到Kubernetes集群中。使用-f选项指定要应用的文件名。

根据输出信息，可以看出已经成功创建了以下资源：

```
configmap/mysql-config created
deployment.apps/envfrom-demo created
```

查看Pod资源信息：

```
[root@k8s-master ~]# kubectl get pods -n dean -l name=envfrom-demo
NAME                              READY   STATUS    RESTARTS   AGE
envfrom-demo-5dfdb6fc48-7c4vc     1/1     Running   0          110s
```

该命令用于查询Kubernetes集群中dean命名空间下所有带有标签name=envfrom-demo的Pods。

查看ConfigMap资源信息：

```
[root@k8s-master ~]# kubectl get configmap -n dean mysql-config
NAME           DATA   AGE
mysql-config   2      2m4s
```

查看Pod内的变量信息，由此可以看到ConfigMap的信息已经注入Pod中：

```
[root@k8s-master ~]# kubectl exec -it -n dean envfrom-demo-5dfdb6fc48-7c4vc -- printenv
# 输出信息
PATH=/usr/local/sbin:/usr/local/bin:/usr/sbin:/usr/bin:/sbin:/bin
HOSTNAME=envfrom-demo-5dfdb6fc48-7c4vc
TERM=xterm
master.cnf=
[mysqld]
datadir=/var/lib/mysql
socket=/var/lib/mysql/mysql.sock
symbolic-links=0
server-id=1
slave.cnf=
[mysqld]
datadir=/var/lib/mysql
socket=/var/lib/mysql/mysql.sock
symbolic-links=0
server-id=2

DEANIT_SVC_EXTERNAL_PORT_80_TCP=tcp://10.105.23.32:80
DEANIT_SVC_EXTERNAL_PORT_443_TCP=tcp://10.105.23.32:443
```

7.1.7 通过 valueFrom 方式指定 ConfigMap

在k8s-master节点创建名为special-config的ConfigMap，其中定义了两个KEY：

```
[root@k8s-master ~]# kubectl create configmap special-config --from-literal=special.how=very --from-literal=special.type=charm -n dean
```

在k8s-master节点的资源清单中创建Pod，并通过valueFrom方式指定ConfigMap：

```
[root@k8s-master ~]# vim valuefrom_demo.yaml
apiVersion: v1
kind: Pod
```

```yaml
metadata:
  name: dapi-cm-pod
  namespace: dean
spec:
  containers:
    - name: test-container
      image: busybox
      command: [ "/bin/sh", "-c", "env" ]
      env:
        - name: SPECIAL_LEVEL_KEY          # 变量名，ConfigMap的键的value作为值
          valueFrom:
            configMapKeyRef:
              name: special-config          # ConfigMap的名字
              key: special.how              # ConfigMap的键
  restartPolicy: Never
```

应用资源：

```
[root@k8s-master ~]# kubectl apply -f valuefrom_demo.yaml
pod/dapi-cm-pod created
```

该命令使用Kubectl工具将valuefrom_demo.yaml文件中定义的资源应用到Kubernetes集群中。使用-f选项指定要应用的文件名。根据输出信息，可以看出已经成功创建了以下资源：

```
pod/dapi-cm-pod created
```

查看ConfigMap资源信息：

```
# 查看ConfigMap
[root@k8s-master ~]# kubectl get cm -n dean | grep special-config
special-config      2       3m20s
```

查看Pod资源信息：

```
[root@k8s-master ~]# kubectl get pods -n dean
NAME               READY   STATUS      RESTARTS   AGE
dapi-cm-pod        0/1     Completed   0          35s
```

7.1.8　Nginx 通过 ConfigMap 管理配置文件

subpath的作用如下。

- 避免覆盖：如果挂载路径是一个已存在的目录，则目录下的内容不会被覆盖，直接将configMap/Secret挂载到容器的路径下，会覆盖掉容器路径下原有的文件，使用subpath选定configMap/Secret中的特定key-value并挂载到容器中，将不会覆盖原目录下的其他文件。
- 文件隔离：Pod中含有多个容器共用一个日志volume，不同容器的日志路径挂载到不同的子目录，而不是根路径（subpath目录会在底层存储自动创建且权限为777，无须手动创建）。

在k8s-master节点创建资源清单：

```
[root@k8s-master ~]# vim nginx_subpath.yaml
apiVersion: v1
kind: ConfigMap
metadata:
  name: nginx-prod
  namespace: dean
data:
    # 多行字符串可以使用 | 保留换行
    # +表示保留文字块末尾的换行
    # -表示删除字符串末尾的换行
  nginx.conf: |-
    user  nginx;
    worker_processes  auto;
    worker_cpu_affinity 00000001 00000010 00000100 00001000;

    error_log  /var/log/nginx/error.log warn;
    pid        /var/run/nginx.pid;


    worker_rlimit_nofile 65536;

    events {
        worker_connections  65535;
        accept_mutex on;
        multi_accept on;
    }
    http {
        include       mime.types;
        default_type  application/octet-stream;
        log_format access_json '{"@timestamp":"$time_iso8601",'
        '"host":"$server_addr",'
        '"clientip":"$remote_addr",'
        '"size":$body_bytes_sent,'
        '"responsetime":$request_time,'
        '"upstreamtime":"$upstream_response_time",'
        '"upstreamhost":"$upstream_addr",'
        '"http_host":"$host",'
        '"url":"$uri",'
        '"domain":"$host",'
        '"xff":"$http_x_forwarded_for",'
        '"referer":"$http_referer",'
        '"status":"$status"}';
        access_log  /var/log/nginx/access.log  access_json;

        client_max_body_size 50M;
        keepalive_timeout  300;
        fastcgi_buffers 8 128k;
        fastcgi_buffer_size 128k;
        fastcgi_busy_buffers_size 256k;
        fastcgi_temp_file_write_size 256k;
```

```
            proxy_connect_timeout 90;
            proxy_read_timeout 300;
            proxy_send_timeout 300;
            sendfile        on;

            server {
                listen      80;
                server_name localhost;
                add_header Cache-Control no-cache;

                location / {
                    root   /usr/share/nginx/html;
                        #proxy_read_timeout 220s
                    index  index.html index.htm;
                }

                error_page   500 502 503 504  /50x.html;
                location = /50x.html {
                    root   html;
                }
            }
            include /etc/nginx/conf.d/*.conf;
        }
---
apiVersion: apps/v1
kind: Deployment
metadata:
  name: nginx-prod
  namespace: dean
spec:
  replicas: 1
  template:
    metadata:
      labels:
        app: nginx-prod
    spec:
      containers:
      - name: nginx
        image: nginx:1.21
        imagePullPolicy: IfNotPresent
        volumeMounts:
        - mountPath: /etc/nginx/nginx.conf
          name: nginx-config
          subPath: etc/nginx/nginx.conf   # 这里要和下面的path保持一致，etc前面不能写/（不能用绝对路径)
        ports:
        - containerPort: 80
      volumes:
      - name: nginx-config
```

```yaml
        configMap:
          name: nginx-prod
          items:
          - key: nginx.conf
            path: etc/nginx/nginx.conf    # 这里要和上面的subpath保持一致，etc前面不能写/（不
能用绝对路径）
      selector:
        matchLabels:
          app: nginx-prod
```

应用资源：

```
[root@k8s-master ~]# kubectl apply -f nginx_subpath.yaml
configmap/nginx-prod created
deployment.apps/nginx-prod created
```

该命令使用Kubectl工具将nginx_subpath.yaml文件中定义的资源应用到Kubernetes集群中。使用-f选项指定要应用的文件名。根据输出信息，可以看出已经成功创建了以下资源：

```
configmap/nginx-prod created
deployment.apps/nginx-prod created
```

查看ConfigMap资源信息：

```
# 查看ConfigMap
[root@k8s-master ~]# kubectl get cm -n dean | grep special-config
nginx-prod       1       31s
```

查看Pod资源信息：

```
[root@k8s-master ~]# kubectl get pods -n dean -l app=nginx-prod
NAME                              READY   STATUS    RESTARTS   AGE
nginx-prod-649c78fc7b-jbtnb       1/1     Running   0          25s
```

查看Pod内的Nginx目录及配置文件信息，由此可见使用subpath可以避免目录覆盖问题：

```
# 查看/etc/nginx目录下的文件
[root@k8s-master ~]# kubectl exec -it -n dean nginx-prod-649c78fc7b-jbtnb -- ls -l /etc/nginx/
total 24
drwxr-xr-x 1 root root   26 Jun 24 06:06 conf.d
-rw-r--r-- 1 root root 1007 Dec 28  2021 fastcgi_params
-rw-r--r-- 1 root root 5349 Dec 28  2021 mime.types
lrwxrwxrwx 1 root root   22 Dec 28  2021 modules -> /usr/lib/nginx/modules
-rw-r--r-- 1 root root 1914 Jun 24 06:06 nginx.conf
-rw-r--r-- 1 root root  636 Dec 28  2021 scgi_params
-rw-r--r-- 1 root root  664 Dec 28  2021 uwsgi_params

# 查看Nginx配置文件的server_name配置
[root@k8s-master ~]# kubectl exec -it -n dean nginx-prod-649c78fc7b-jbtnb -- cat /etc/nginx/nginx.conf | grep server_name  server_name  localhost;
```

7.2 Secret：敏感信息的安全存储与访问

本节将详细介绍Secret的基本概念、用途、创建方式以及使用场景。

7.2.1 Secret 概述

Secret是Kubernetes中用于存储敏感信息的资源类型，如密码、密钥、OAuth令牌等。与ConfigMap类似，Secret也以键-值对的形式存储数据，但它是加密的，以确保敏感信息的安全性。

Secret能够以Volume或者环境变量的方式使用，Secret如果是以subPath的形式挂载的，那么Pod是不会感知到Secret的更新的。如果Pod的变量来自Secret中定义的内容，那么Secret更新后也不会更新Pod中的变量。

Secret的主要用途包括：

- Secret可以作为环境变量的来源，供Pod中的容器使用敏感信息（如数据库密码）。
- Secret可以挂载为卷，供容器中的进程读取敏感信息（如SSL证书）。

Secret的创建方式与ConfigMap类似，可以通过Kubectl工具或YAML文件创建。但需要注意的是，在创建Secret时，需要指定一个类型，以标识Secret的用途和加密方式，类型如下：

- generic：通用类型，通常用于存储密码数据。
- tls：此类型仅用于存储私钥和证书。
- docker-registry：若保存Docker仓库的认证信息，则必须使用这种类型来创建。

Secret的使用场景：

- 当应用程序需要访问敏感信息（如数据库密码、API密钥）时，可以使用Secret来存储和管理这些信息。
- 当需要在多个Pod之间共享敏感信息时，可以使用Secret来避免在多个Pod中重复配置。

7.2.2 使用文件方式创建 Secret

使用echo命令将账号和密码通过>输出到对应文件中：

```
[root@k8s-master ~]# echo "admin" >/root/username.txt
[root@k8s-master ~]# echo "dean" >/root/password.txt
```

基于文件创建Secret：

```
[root@k8s-master ~]# kubectl create secret generic db-user-pass --from-file=/root/username.txt --from-file=/root/password.txt -n dean
secret/db-user-pass created
```

该命令使用Kubectl工具在Kubernetes集群中创建了一个名为db-user-pass的通用（generic）类型的Secret。这个Secret用于存储数据库的用户名和密码，这些用户名和密码分别存储在本地文件/root/username.txt和/root/password.txt中。-n dean参数指定了这个Secret将在名为dean的命名空间中创建。根据输出信息，可以看出已经成功创建了以下资源：

```
secret/db-user-pass created
```

查看Kubernetes集群中dean命名空间下名为db-user-pass的Secret并以YAML格式输出：

```
[root@k8s-master ~]# kubectl get secrets -n dean db-user-pass -o yaml
apiVersion: v1
data:
  password.txt: ZGVhbgo=
  username.txt: YWRtaW4K
kind: Secret
metadata:
  creationTimestamp: "2024-06-04T15:38:53Z"
  managedFields:
  - apiVersion: v1
    fieldsType: FieldsV1
    fieldsV1:
      f:data:
        .: {}
        f:password.txt: {}
        f:username.txt: {}
      f:type: {}
    manager: kubectl-create
    operation: Update
    time: "2024-06-04T15:38:53Z"
  name: db-user-pass
  namespace: dean
  resourceVersion: "9699512"
  selfLink: /api/v1/namespaces/default/secrets/db-user-pass
  uid: a24ced1d-2f3f-4917-8012-96c580659ee6
# base64编码格式的Secret，用来存储密码、密钥等
type: Opaque
```

对上述Secret中的密文使用base64 -d命令进行解密，可以看到最初设置的账号和密码：

```
[root@k8s-master ~]# echo "YWRtaW4K" | base64 -d
admin
[root@k8s-master ~]# echo "ZGVhbgo=" | base64 -d
dean
```

该命令是对base64加密后的密文值进行解密。

- base64编码使用64个字符的集合来表示二进制数据。这个集合包括大写字母A～Z、小写字母a～z、数字0～9、加号（+）、斜杠（/），以及一个用于填充的等号=（当原始数据长度

不是3的倍数时）。在大多数UNIX-like系统（如Linux和macOS）中，base64是一个命令行工具，用于对文件或标准输入进行base64编码或解码。
- base64 -d和base64 --decode选项都是用来指定base64命令进行解码操作的。-d是短选项，而--decode是长选项，它们在功能上是等价的。选择哪一个主要取决于个人喜好或脚本的编写习惯。

7.2.3 使用 YAML 方式创建 Secret

使用base64工具生成用户名和密码的密文：

```
[root@k8s-master ~]# echo "admin" | base64
YWRtaW4K
[root@k8s-master ~]# echo "123456" | base64
MTIzNDU2Cg==
```

资源清单示例：

```
apiVersion: v1
kind: Secret
metadata:
  name: mysecret
  namespace: dean
type: Opaque
data:
  # 使用通过base64工具生成的密文
  username: YWRtaW4=
  password: MTIzNDU2Cg==
```

7.2.4 Secret 权限解析

Secret具有权限模式位。如果不指定，则默认使用0644（八进制），即420（十进制）。

- YAML：可以使用0644、0755、0777等八进制权限值，前面一定要写0。
- JSON：不支持八进制表示法，应使用十进制。例如256（十进制）等于0400（八进制），511（十进制）等同于0777（八进制）。

7.2.5 使用 Docker 的 config.json 方式创建 Secret

使用docker login登录镜像仓库，此时在本地/root/.docker目录下会生成一个config.json文件，config.json文件包含登录镜像仓库的认证信息：

```
[root@k8s-master ~]# docker login https://harbor.deanit.cn
```

将config.json文件使用base64工具进行加密，执行以下命令：

```
[root@k8s-master ~]# base64 -w 0 /root/.docker/config.json
```

该命令用于将/root/.docker/config.json文件的内容进行base64编码，并且不使用任何换行符（-w 0参数指定了输出宽度为0，即不自动换行）。

资源清单示例：

```
apiVersion: v1
kind: Secret
metadata:
  name: image-pull-secret
  namespace: dean
data:
  .dockerconfigjson: >-
    eyJhdXRocyXIOXN【<---此行替换成上面由base64工具生成的密文--->】YXNleWN1aWIifX19
type: kubernetes.io/dockerconfigjson
```

7.2.6 使用 Kubectl 创建 Docker Registry 认证的 Secret

使用Kubectl工具创建一个Docker Registry的Secret：

```
[root@k8s-master ~]# kubectl create secret -n dean docker-registry harbor-key --docker-server=harbor.deanit.cn --docker-username=admin --docker-password=123456 --docker-email=youremail@qq.com
```

参数解析：

- docker-registry：Secret类型。
- harbor-key：Secret名字。
- --docker-server：镜像仓库的地址。
- --docker-username：用户名。
- --docker-password：密码。
- --docker-email：邮箱。

资源清单示例：

```
apiVersion: v1
kind: Pod
metadata:
  name: dean-pod
  namespace: dean
spec:
  containers:
    - name: dean
      image: harbor.deanit.cn/myapp-nginx/myapp-nginx:v1
      ports:
        - containerPort: 80
  # 镜像拉取Secret
  imagePullSecrets:
```

```
    # Secret的名字
    - name: harbor-key
```

7.3 本章小结

通过本章的学习，读者将能够熟练掌握在Kubernetes集群中使用ConfigMap和Secret。通过它们提供灵活、安全的存储和访问机制，帮助开发人员更好地管理和保护应用程序的配置和敏感信息。无论是初学者还是经验丰富的Kubernetes运维人员，都能从中获得宝贵的知识和经验。

第 8 章

Kubernetes鉴权机制

在云计算和容器化技术飞速发展的今天，Kubernetes凭借其出色的容器编排和集群管理能力，已经成为云原生应用的首选平台。然而，随着集群规模的扩大和复杂性的增加，如何确保集群的安全性和访问控制成为至关重要的问题。本章将深入探讨Kubernetes的鉴权机制，为读者提供全面的理解和认识。

8.1 Kubernetes鉴权机制概述

Kubernetes的鉴权机制主要基于两个核心概念：认证（Authentication）和授权（Authorization）。

（1）认证：验证用户或实体的身份是否合法。Kubernetes支持多种认证方式，包括基于证书、基于令牌、基于HTTP基本认证等。通过认证，系统可以确定请求的来源是否可信。

（2）授权：根据认证结果，判断用户或实体是否有权访问特定的资源。Kubernetes的授权机制基于RBAC（Role-Based Access Control，基于角色的访问控制）模型，通过定义角色（Role）和角色绑定（RoleBinding），将权限与用户或实体关联起来。同时，Kubernetes还支持基于属性的访问控制（Attribute-Based Access Control，ABAC）和WebHook等扩展授权方式，以满足不同的安全需求。

8.2 鉴权机制的工作流程

在Kubernetes中，鉴权机制的工作流程大致如下。

（1）客户端发起请求：用户或实体通过API Server向Kubernetes集群发起请求，请求访问特定的资源。

（2）认证阶段：APIServer对请求进行认证，验证用户或实体的身份是否合法。如果认证失败，则拒绝请求并返回错误信息。

（3）授权阶段：如果认证成功，APIServer将进入授权阶段。在这一阶段，系统将根据请求中的资源、操作和用户或实体的身份信息，查询相应的角色和角色绑定，以确定用户或实体是否有权执行该操作。如果授权失败，则拒绝请求并返回错误信息。

（4）访问控制决策：基于认证和授权的结果，APIServer将做出访问控制决策。如果请求被允许，则APIServer将执行相应的操作并返回结果；否则，将拒绝请求并返回错误信息。

8.3 角色/角色绑定概述

在Kubernetes中，RBAC（Role-Based Access Control，基于角色的访问控制）是一个重要的安全特性，它允许管理员通过定义角色（Role）和角色绑定（RoleBinding）来管理集群的访问权限。这里的角色与角色绑定涉及以下概念。

1. Role（角色）

- 定义：Role是一组权限的集合，它允许对特定命名空间中的Kubernetes资源进行访问。
- 范围：Role的作用范围限定在单个命名空间内。
- 示例：一个Role可以定义在default命名空间中，用于授予对该命名空间中Pods的读取权限。

2. ClusterRole（集群角色）

- 定义：ClusterRole与Role类似，也是一组权限的集合，但ClusterRole的作用范围是整个集群，包括所有命名空间和集群级别的资源。
- 范围：ClusterRole不受命名空间的限制，可以用于控制集群级别的资源、非资源端点以及跨命名空间的资源访问。
- 示例：一个ClusterRole可以授予对集群中所有Pods的读取权限，或者对特定命名空间下的Secrets的读取权限。

3. RoleBinding（角色绑定）

- 定义：RoleBinding用于将Role绑定到一个或多个用户（User）、用户组（Group）或服务账户（ServiceAccount），从而授予这些实体对Role中定义的资源的访问权限。
- 范围：RoleBinding的作用范围限定在单个命名空间内。
- 示例：一个RoleBinding可以将一个Role绑定到一个特定的ServiceAccount，从而允许该ServiceAccount在指定的命名空间中执行Role中定义的权限。

4. ClusterRoleBinding（集群角色绑定）

- 定义：ClusterRoleBinding用于将ClusterRole绑定到一个或多个用户（User）、用户组（Group）或服务账户（ServiceAccount），从而授予这些实体对ClusterRole中定义的资源的访问权限。

- 范围：ClusterRoleBinding的作用范围是整个集群，不受命名空间的限制。
- 示例：一个ClusterRoleBinding可以将一个ClusterRole绑定到一个特定的ServiceAccount，从而允许该ServiceAccount在整个集群中执行ClusterRole中定义的权限。

8.4 用户鉴权实战

本实例将详细介绍如何配置Kubernetes集群，以确保仅dean用户能够操作集群中名为dean的命名空间内的资源。通过精细的RBAC策略设置，我们可以有效地限制dean用户的权限范围，仅授权其对dean命名空间内的资源进行管理操作，从而增强集群的安全性和管理的灵活性。

具体操作步骤如下：

01 创建私钥：

```
[root@k8s-master ~]# cd /etc/kubernetes/pki/
[root@k8s-master pki]# (umask 077; openssl genrsa -out dean.key 2048)
# 输出信息
Generating RSA private key, 2048 bit long modulus
..............................................................................+++
..............................................................................+++
e is 65537 (0x10001)
```

该命令用于在保护新生成私钥文件（dean.key）的安全性的同时，生成一个2048位的RSA私钥。通过设置umask 077，确保了只有文件的创建者（在本例中为执行命令的用户）可以访问该文件，从而增强了安全性。

参数解析：

- umask：是一个系统调用，用于设置创建新文件或目录时的默认权限掩码。
- 077：表示将文件或目录的默认权限设置为仅所有者可读写执行（对于目录，执行权限表示可以进入目录）。
- openssl：是一个强大的安全套接字层（Secure Sockets Layer，SSL）和传输层安全（Transport Layer Security，TLS）协议工具，以及一个通用加密库。
- genrsa：是OpenSSL中用于生成RSA私钥的命令。
- -out dean.key：指定了输出文件的名称。
- 2048：指定了生成的RSA密钥的长度为2048位。这是目前推荐的长度之一，以提供足够的安全性。

02 创建证书签署请求，-subj选项中的CN的值将被kubeconfig作为用户名使用，O的值将被识别为用户组：

```
[root@k8s-master pki]# openssl req -new -key dean.key -out dean.csr -subj "/CN=dean/O=kubernetes"
```

参数解析：

- openssl req：这是OpenSSL中用于处理证书签名请求的命令。
- -new：指示OpenSSL生成一个新的证书签名请求。
- -key dean.key：指定用于生成证书签名请求的私钥文件。在这个例子中，私钥文件名为dean.key，它应该是一个之前已经生成的RSA私钥文件。
- -out dean.csr：指定输出文件的名称，即生成的证书签名请求文件。在这个例子中，输出文件名为dean.csr。
- -subj "/CN=dean/O=kubernetes"：允许在命令行中直接指定证书的主题（Subject）信息，而不是在生成过程中通过交互式提示输入。/CN=dean/O=kubernetes表示证书的主题包含公用名（Common Name，CN）为dean和组织名（Organization，O）为kubernetes。这些信息将包含在生成的证书签名请求中，并最终可能出现在由证书颁发机构（Certificate Authority，CA）签发的证书中。

03 基于Kubeadm安装Kubernetes集群时生成的CA签署证书，设置有效时长为3650天：

```
[root@k8s-master pki]# openssl x509 -req -in dean.csr -CA ca.crt -CAkey ca.key -CAcreateserial -out dean.crt -days 3650
# 输出信息：
Signature ok
subject=/CN=dean/O=kubernetes
Getting CA Private Key
```

参数解析：

- openssl x509：这是OpenSSL中用于处理x509证书的命令。
- -req：指示OpenSSL该命令将处理一个证书签名请求。
- -in dean.csr：指定输入文件，即要签名的证书签名请求文件。
- -CA ca.crt：指定证书颁发机构的证书文件。这个证书用于验证CA的身份，并用于在签发的证书中嵌入CA的信息。
- -CAkey ca.key：指定CA的私钥文件。这个私钥用于对CSR进行签名，以生成有效的证书。
- -CAcreateserial：如果在CA的目录中不存在序列号文件（通常是serial或serial.txt），则自动创建一个新的序列号文件，并生成一个新的序列号用于新签发的证书。
- -out dean.crt：指定输出文件的名称，即新生成的、已签名的证书文件。
- -days 3650：指定证书的有效期，这里是3650天，即大约10年。

04 查看证书内容：

```
[root@k8s-master pki]# openssl x509 -in dean.crt -text -noout
Certificate:
    Data:
        Version: 1 (0x0)
        Serial Number:
```

```
            ee:f8:72:bb:cb:b1:2f:7f
        Signature Algorithm: sha256WithRSAEncryption
            Issuer: CN=kubernetes
            Validity
                Not Before: Jun 25 03:18:11 2024 GMT
                Not After : Jun 23 03:18:11 2034 GMT
            Subject: CN=dean, O=kubernetes
            Subject Public Key Info:
                Public Key Algorithm: rsaEncryption
                    Public-Key: (2048 bit)
                    Modulus:
                        00:9e:42:c1:29:39:86:f0:09:87:85:03:f5:67:8b:
                        12:3f:95:8a:b7:12:6a:24:56:d6:72:f8:da:47:82:
                        28:22:ab:3d:55:0f:bd:9a:8c:da:51:40:79:2e:35:
                        3a:e2:ae:e1:e1:a6:6e:36:72:a2:36:73:6c:e3:ad:
                        d3:fa:e9:13:d2:ff:a8:d0:52:c9:dc:5c:3f:f7:db:
                        bb:28:69:a2:b5:33:98:e4:0f:5b:fe:61:31:58:14:
                        57:e6:09:78:2e:01:e4:6c:15:33:9a:f3:ed:18:33:
                        fb:f1:fc:08:a0:0f:5e:ef:a8:86:77:b0:bc:7a:dc:
                        95:47:47:e8:53:17:24:bd:b2:56:8e:0f:ba:2f:ba:
                        ba:d4:71:86:87:43:9d:d7:3f:61:ff:f2:e9:d9:58:
                        b0:d2:75:c2:3c:a1:f9:0a:48:83:60:eb:f9:fa:79:
                        29:3b:33:91:c7:41:c3:b5:02:27:e4:db:38:8f:52:
                        a9:67:70:ce:3d:aa:09:64:53:05:38:8e:e5:65:a4:
                        c7:ab:5f:fb:00:27:23:a8:4b:a5:c2:cb:80:94:7a:
                        b6:63:6f:cd:89:b7:03:03:ed:2a:64:7f:06:ff:67:
                        75:82:9b:91:7f:e8:6e:58:28:a9:53:96:f7:8b:16:
                        eb:03:2d:bf:18:e0:1c:f3:6b:a0:ec:86:7f:b8:15:
                        ed:d1
                    Exponent: 65537 (0x10001)
        Signature Algorithm: sha256WithRSAEncryption
            50:10:a5:f2:7d:5b:43:68:22:fc:e7:56:52:74:71:c3:c8:f5:
            5b:e5:b1:41:a6:b3:0c:1e:79:dc:e5:a2:ed:84:e9:e8:13:0f:
            f5:1c:6b:5e:bb:22:65:fb:20:8b:c8:05:0a:40:45:53:a6:c4:
            2b:0e:a8:d9:62:22:d6:19:6b:3a:be:cc:b8:1e:27:6a:7d:91:
            7b:25:88:dc:ee:ce:84:ba:6d:05:03:e2:12:fe:42:ba:f9:90:
            e6:c1:d3:4a:d8:38:5f:f8:72:30:bb:b8:45:a4:56:8d:df:6f:
            a6:0e:65:f3:93:45:21:9e:0d:9d:ce:25:a4:a9:4a:b2:83:89:
            a5:28:80:f1:03:71:d6:26:d1:f2:03:bd:1e:d8:72:88:c3:be:
            d7:ae:9b:c1:3c:df:65:de:c3:35:08:3b:b4:59:61:cc:37:da:
            c8:2f:9b:23:cf:96:9f:38:93:ac:3a:79:77:29:13:c6:00:91:
            27:c4:39:1c:b4:6c:eb:9b:32:3a:ba:39:e6:82:b9:0b:ec:3b:
            0b:33:49:10:a1:3c:fa:df:43:ec:d5:c4:9b:84:de:56:34:bd:
            0c:dc:70:d3:77:76:7b:2f:63:3a:60:cf:20:73:81:8b:53:e9:
            af:79:fd:15:f9:e0:e3:25:d6:55:b3:86:29:42:18:9f:e8:8c:
            77:f1:e8:cc
```

执行该命令后,你会在终端或命令行界面中看到dean.crt证书的详细文本信息,包括证书的所有字段和值。这对于检查证书的有效性、确认证书的颁发者和主体信息、验证证书的签名和公钥等操作非常有用。

05 将dean这个用户添加到Kubernetes集群中,可以用来认证APIServer的连接:

```
[root@k8s-master pki]# kubectl config set-credentials dean
--client-certificate=./dean.crt --client-key=./dean.key --embed-certs=true
# 输出信息
User "dean" set.
```

该命令用于配置Kubernetes客户端(Kubectl)的凭据。这个命令会更新Kubernetes的配置文件(通常是 ~/.kube/config),为名为dean的用户设置客户端证书和私钥。这里,--client-certificate=./dean.crt指定了客户端证书的路径,--client-key=./dean.key指定了对应的私钥文件的路径。--embed-certs=true是一个可选参数,它指示Kubectl将证书和私钥的内容直接嵌入配置文件中,而不是仅仅引用它们的文件路径。这样做的好处是配置文件变得更加便携,因为它不依赖于外部文件的存在。

06 在kubeconfig下新增dean账号,配置context,用来组合cluster和credentials,即访问的集群的上下文:

```
[root@k8s-master pki]# kubectl config set-context dean@kubernetes
--cluster=kubernetes --user=dean
# 输出信息
Context "dean@kubernetes" created.
```

执行该命令后,dean@kubernetes上下文将被设置或更新,以使用Kubernetes集群和dean用户。但是,这个上下文不会自动成为当前上下文。要使其成为当前上下文,需要使用kubectl config use-context dean@kubernetes命令,执行此命令后,dean@kubernetes将成为当前上下文,所有后续的kubectl命令都将使用这个上下文来与Kubernetes集群交互。

07 在k8s-master节点创建角色资源清单:

```
[root@k8s-master ~]# vim role.yaml
kind: Role
apiVersion: rbac.authorization.k8s.io/v1
metadata:
  namespace: dean        # 命名空间
  name: dean             # role名字
rules:
- apiGroups: [""]
  # 指定的角色或用户拥有对核心组中Pods的访问权限,包括查看和操作Pods,以及执行对Pods的exec命令
  resources: ["pods","pods/exec"]
  # 可以操作Pod的动作
  verbs: ["get","logs","watch","list","create","update","patch","delete"]
- apiGroups: [""]
  # 可操作svc
  resources: ["services"]
  # 可以操作svc的动作
  verbs: ["get","watch","list"]
```

08 在k8s-master节点创建角色绑定资源清单：

```yaml
[root@k8s-master ~]# vim rolebinding.yaml
kind: RoleBinding
apiVersion: rbac.authorization.k8s.io/v1
metadata:
  name: dean                    # RoleBinding的名称
  namespace: dean                # 命名空间
subjects:
  - kind: User
    name: dean
    apiGroup: rbac.authorization.k8s.io
roleRef:
  kind: Role                     # 类型为Role
  name: dean                     # 指定上面创建的Role名称
  apiGroup: rbac.authorization.k8s.io
```

09 应用资源：

```
[root@k8s-master ~]# kubectl create -f role.yaml
role.rbac.authorization.k8s.io/dean created
[root@k8s-master ~]# kubectl create -f rolebinding.yaml
rolebinding.rbac.authorization.k8s.io/dean created
```

该命令使用Kubectl工具将role.yaml、rolebinding.yaml文件中定义的资源应用到Kubernetes集群中。使用-f选项指定要应用的文件名。根据输出信息，可以看出已经成功创建了以下资源：

- role.rbac.authorization.k8s.io/dean created
- rolebinding.rbac.authorization.k8s.io/dean created

10 切换集群用户上下文，进行验证权限：

```
--- 切换到dean上下文
[root@k8s-master ~]# kubectl config use-context dean@kubernetes
Switched to context "dean@kubernetes".

--- 查看当前上下文
[root@k8s-master ~]# kubectl config get-contexts
CURRENT   NAME                              CLUSTER      AUTHINFO            NAMESPACE
*         dean@kubernetes                   kubernetes   dean
          kubernetes-admin@kubernetes       kubernetes   kubernetes-admin

--- 查看default命名空间下的资源，通过输出信息可以发现没有权限
[root@k8s-master ~]# kubectl get pods
Error from server (Forbidden): pods is forbidden: User "dean" cannot list resource "pods" in API group "" in the namespace "default"

--- 查看dean命名空间下的资源，可以正常显示
[root@k8s-master ~]# kubectl get pods -n dean
NAME                              READY   STATUS     RESTARTS        AGE
```

```
envfrom-demo-5dfdb6fc48-7c4vc        1/1     Running     0            25h
```

11 在k8s-master节点创建一个名为dean的普通用户，将集群配置文件复制到dean用户的.kube目录中：

```
--- 使用useradd命令创建用户
[root@k8s-master ~]# useradd dean

--- 复制root下的.kube目录到dean用户下
[root@k8s-master ~]# cp -ar /root/.kube/ /home/dean/

--- 设置dean用户的所属组和所属用户
[root@k8s-master ~]# chown -R dean.dean /home/dean/

--- 切换到dean用户
[root@k8s-master ~]# su - dean

--- 执行查看default命名空间下的Pod资源信息
[dean@k8s-master ~]$ kubectl get pods
Error from server (Forbidden): pods is forbidden: User "dean" cannot list resource "pods" in API group "" in the namespace "default"

--- 查看dean命名空间下的资源，可以正常显示
[dean@k8s-master ~]$ kubectl get pods  -n dean
NAME                              READY   STATUS    RESTARTS    AGE
envfrom-demo-5dfdb6fc48-7c4vc     1/1     Running   0           25h
```

12 修改dean用户.kube目录下的config文件：

```
[dean@k8s-master ~]$ vim .kube/config
```

13 将图8-1中框中关于kubernetes-admin的集群信息删除并保存。

图 8-1　删除 K8s 集群管理员信息

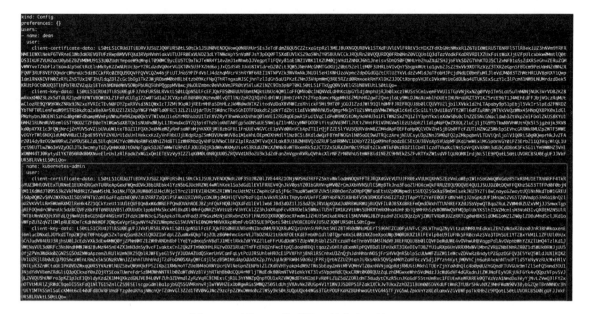

图 8-1 删除 K8s 集群管理员信息（续）

运维笔记：如果你在一个多用户环境中管理Kubernetes集群，并且想要提高集群的安全性，删除或限制非管理员用户对kubernetes-admin用户或集群配置的访问是一个合理的做法。我们将图8-1中框中的内容删除，这样其他用户将只能查看分配的资源权限。

8.5 本章小结

通过本章的学习，读者将能够深入地理解和掌握在Kubernetes集群中鉴权机制的重要性与实现方法。Kubernetes作为一个高度可扩展的容器编排平台，其安全性与访问控制是保障集群稳定运行和数据安全的关键。本章详细阐述了Kubernetes的鉴权体系，包括认证（Authentication）和授权（Authorization）两大核心环节，以及它们如何协同工作以确保只有经过授权的用户或系统能够访问集群资源。无论是初学者还是经验丰富的Kubernetes运维人员，都能从中获得宝贵的知识和经验。

第 9 章

容器运行时Containerd

 从Kubernetes v1.24起,Kubernetes默认使用Containerd替代Docker作为容器运行时,Containerd的用法和Docker类似,有Docker基础的读者转型使用Containerd也相当简单。本章将为读者介绍Containerd的基本概念、安装配置与使用。

9.1 Containerd概述

Containerd是一个由Docker团队开发并开源的、高度可扩展的容器运行时。它负责管理容器的生命周期、镜像管理以及资源隔离等核心功能。作为云原生计算基金会(Cloud Native Computing Foundation,CNCF)的顶级项目之一,Containerd已经成为现代容器化部署和管理的关键工具。Containerd是一个行业标准的容器运行时,强调简单性、健壮性和可移植性。它可作为Linux和Windows的守护进程使用,可以管理其主机系统的完整容器生命周期,如图像传输和存储、容器执行和监督、低级存储和网络附件等。Containerd旨在嵌入更大的系统中,而不是由开发人员或最终用户直接使用,Containerd底层架构如图9-1所示。

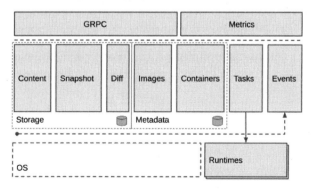

图9-1 Containerd 底层架构图

9.2 安装与配置Containerd

本节将通过一系列简明易懂的步骤，带领读者安装并配置Containerd。安装Containerd不仅能够增强读者对容器运行时技术的理解，还能够为后续深入学习Kubernetes、Docker等容器编排与容器化技术打下坚实的基础。通过亲手操作，读者将亲身体验到Containerd如何高效地管理容器的生命周期，包括容器的创建、运行、停止和删除等关键环节。

9.2.1 安装 Containerd

我们将介绍在CentOS 7系统上安装Containerd的具体方法，推荐计算机最低使用2核CPU，4G内存配置。具体安装步骤如下：

01 下载Docker阿里源并生成缓存：

```
[root@containerd ~]# wget -O /etc/yum.repos.d/docker-ce.repo https://mirrors.aliyun.com/docker-ce/linux/centos/docker-ce.repo && yum makecache fast
```

这两条命令将使用阿里云镜像站点提供的Docker CE软件源，从而加速Docker CE软件包的下载速度，并重新生成缓存的软件包和元数据。具体来说，它执行以下操作：

- 使用wget命令从阿里云镜像站点下载Docker CE的软件源配置文件，并将其保存到/etc/yum.repos.d/目录下的docker-ce.repo文件中。
- 使用yum makecache fast命令重新生成缓存的软件包和元数据，以便系统能够正确识别新的软件源。

02 安装Containerd软件包：

```
[root@containerd ~]# yum install -y containerd.io
```

03 创建Containerd配置文件存放目录并生成配置文件：

```
[root@containerd ~]# mkdir /etc/containerd -p
[root@containerd ~]# containerd config default > /etc/containerd/config.toml
```

执行这条命令后，/etc/containerd/config.toml文件将被填充为Containerd的默认配置。如果需要修改Containerd的配置（比如更改镜像存储位置、调整日志级别等），可以直接编辑这个文件，然后重启Containerd服务以使更改生效。

04 编辑Containerd配置文件：

```
--- 将registry.k8s.io镜像地址替换为阿里云镜像
[root@containerd ~]# sed -i "s#registry.k8s.io/pause#registry.aliyuncs.com/google_containers/pause#g" /etc/containerd/config.toml
```

--- 开启systemd控制Cgroup
```
[root@containerd ~]# sed -i 's#SystemdCgroup = false#SystemdCgroup = true#g' /etc/containerd/config.toml
```

--- 在registry.mirrors下加入docker.io的镜像源
```
[root@containerd ~]# sed -i '/registry.mirrors/a\ \ \ \ \ \ [plugins."io.containerd.grpc.v1.cri".registry.mirrors."docker.io"]' /etc/containerd/config.toml
```

--- 在docker.io的镜像源下面加入163的镜像源
```
[root@containerd ~]# sed -i '/registry.mirrors."docker.io"]/a\ \ \ \ \ \ \ \ endpoint = ["http://hub-mirror.c.163.com"]' /etc/containerd/config.toml
```

--- 在163的镜像源下面加入一条registry.k8s.io镜像源
```
[root@containerd ~]# sed -i '/hub-mirror.c.163.com"]/a\ \ \ \ \ \ [plugins."io.containerd.grpc.v1.cri".registry.mirrors."registry.k8s.io"]' /etc/containerd/config.toml
```

--- 在registry.k8s.io镜像源下面加入一条阿里云的镜像源
```
[root@containerd ~]# sed -i '/"registry.k8s.io"]/a\ \ \ \ \ \ \ \ endpoint = ["http://registry.aliyuncs.com/google_containers"]' /etc/containerd/config.toml
```

执行上述命令，使用sed工具在文件内查找并替换指定的内容。-i选项表示直接修改文件内容，而不是输出到标准输出。

05 设置Containerd为开机自启并立即启动：

```
[root@containerd ~]# systemctl enable containerd --now
```

06 查看Containerd服务状态，确保为running：

```
[root@containerd ~]# systemctl status containerd
● containerd.service - containerd container runtime
   Loaded: loaded (/usr/lib/systemd/system/containerd.service; enabled; vendor preset: disabled)
   Active: active (running) since Sun 2024-07-07 14:45:08 CST; 11s ago
     Docs: https://containerd.io
  Process: 2077 ExecStartPre=/sbin/modprobe overlay (code=exited, status=0/SUCCESS)
 Main PID: 2080 (containerd)
    Tasks: 9
   Memory: 14.8M
   CGroup: /system.slice/containerd.service
           └─2080 /usr/bin/containerd
...
Jul 07 14:45:08 containerd containerd[2080]: time="2024-07-07T14:45:08.903205099+08:00" level=info msg="containerd successfully booted in 0.099498s"
```

9.2.2 配置 Containerd 阿里云镜像加速器

为了加速Containerd镜像的下载速度,我们需要修改Containerd的配置文件,添加阿里云镜像加速器地址。请按照以下步骤操作:

01 使用vim编辑器打开Containerd的配置文件:

```
[root@containerd ~]# vim /etc/containerd/config.toml
```

02 修改Containerd配置,添加阿里云镜像加速器地址,在文件末尾添加以下内容:

```
[plugins]
  [plugins.cri.registry.mirrors."docker.io"]
    endpoint = ["https://nty7c4os.mirror.aliyuncs.com"]
```

03 保存并退出vim编辑器。重启Containerd服务,以使更改生效:

```
[root@containerd ~]# systemctl restart containerd
```

9.2.3 配置 Containerd 使用自建镜像仓库

为了使用自建镜像仓库,我们需要修改Containerd的配置文件,添加自建镜像仓库地址。请按照以下步骤操作:

01 使用vim编辑器打开Containerd的配置文件,修改Containerd配置,添加自建镜像仓库地址:

```
[root@containerd ~]# vim /etc/containerd/config.toml
[plugins."io.containerd.grpc.v1.cri".registry.configs]
# harbor.deanit.cn为自建镜像仓库的地址,可以是域名,也可以是IP地址
[plugins."io.containerd.grpc.v1.cri".registry.configs."harbor.deanit.cn".tls]
    # 建立HTTPS连接时跳过对服务器证书的验证
    insecure_skip_verify = true
# harbor.deanit.cn为自建镜像仓库的地址,可以是域名,也可以是IP地址
[plugins."io.containerd.grpc.v1.cri".registry.configs."harbor.deanit.cn".auth]
    # 自建镜像仓库的用户名
    username = "admin"
    # 自建镜像仓库的密码
    password = "Harbor12345"
```

运维笔记:请将harbor.deanit.cn替换为自建的镜像仓库地址。

02 保存并退出vim编辑器。重启Containerd服务,以使配置生效:

```
[root@containerd ~]# systemctl restart containerd
```

9.3 使用nerdctl管理Containerd

Containerd是一个开源的容器运行时,它提供了一种标准化的方式来运行和管理容器。然而,Containerd自带的命令行工具ctr用户体验不佳,因此官方推出了一个新的命令行客户端工具nerdctl,以提供更好的使用体验。nerdctl与Docker的命令行语法大部分兼容,用法类似,使得用户可以无缝切换到nerdctl。

9.3.1 安装 nerdctl

在使用nerdctl之前,首先要安装它,可以按照以下步骤操作:

01 下载nerdctl软件安装包并解压:

```
--- 下载软件包
[root@containerd ~]# wget https://github.com/containerd/nerdctl/releases/download/v1.7.6/nerdctl-1.7.6-linux-amd64.tar.gz
--- 解压软件包至/usr/bin目录
[root@containerd ~]# tar -xf nerdctl-1.7.6-linux-amd64.tar.gz -C /usr/bin/
```

这两条命令用于下载nertctl工具包到当前目录,将解压的内容直接放入/usr/bin/目录。x表示解包(不是解压缩),f表示使用文件,-C用于指定压缩包要解压到的目录。

02 配置nerdctl命令的Tab键自动补全功能,安装所需的依赖包:

```
[root@containerd ~]# yum install -y epel-release bash-completion
```

在Linux系统中,使用yum install -y epel-release bash-completion命令的目的是安装两个重要的软件包:epel-release和bash-completion。这两个软件包主要用于增强系统的功能和用户体验。下面分别解释这两个软件包的作用:

- EPEL(Extra Packages for Enterprise Linux)是一个为Enterprise Linux(如RHEL、CentOS等)提供额外软件包的仓库。这些软件包通常不会包含在标准的发行版仓库中。
- bash-completion是一个Bash shell的自动补全工具,它可以让用户在命令行中输入命令或文件路径时,通过Tab键自动补全命令或路径,极大地提高了命令行操作的效率和准确性。它为许多常用的命令行工具(如yum、git、ssh等)提供了补全支持,使得用户可以更快地输入命令,并减少因拼写错误导致的错误。

使之立即生效:

```
[root@containerd ~]# source /usr/share/bash-completion/bash_completion
```

该命令的作用是加载Bash补全功能,使用source命令(或者等效的.命令)来执行这个脚本。

source命令会在当前shell环境中执行指定的脚本文件，而不是启动一个新的shell来执行。这意味着脚本中定义的变量、函数等都会在当前shell环境中生效。

将命令添加到profile文件并使之立即生效：

```
[root@containerd ~]# echo "source <(nerdctl completion bash)" >>/etc/profile
[root@containerd ~]# source /etc/profile
```

该命令的目的是将nerdctl的Bash补全功能添加到/etc/profile文件中，以便在Bash shell启动时自动加载nerdctl的补全脚本。

03 进行nerdctl命令的Tab键自动补全测试，通过输出信息可以看出命令能够自动提示：

```
[root@containerd ~]# nerdctl r
rename    (rename a container)
restart   (Restart one or more running containers)
rmi       (Remove one or more images)
rm        (Remove one or more containers)
```

9.3.2 nerdctl 常用命令示例

拉取Nginx镜像：

```
[root@containerd ~]# nerdctl pull nginx
docker.io/library/nginx:latest:     esolved        |++++++++++++++++++++++++++++++|
index-sha256:67682...769fae1cc:     done           |++++++++++++++++++++++++++++++|
manifest-sha256:db...9f40979ce:     done           |++++++++++++++++++++++++++++++|
config-sha256:ffff...0d343cbcb:     done           |++++++++++++++++++++++++++++++|
elapsed: 30.9s                      total: 54.1 M (1.8 MiB/s)
```

查看镜像：

```
[root@containerd ~]# nerdctl images
REPOSITORY   TAG      IMAGE ID      CREATED          PLATFORM      SIZE
nginx        latest   0d17b565c37b  40 seconds ago   linux/amd64   146.2 MiB
```

使用Nginx镜像创建一个容器，并进行端口映射：将容器内部的80端口映射到宿主机的80端口：

```
[root@containerd ~]# nerdctl run -d --name=Dean -p 80:80 nginx
5d1c3f1ca77f17d77f75895c29b6ce491d1921e61ad4f59f4ace3046436d10d4
```

查看容器：

```
[root@containerd ~]# nerdctl ps
CONTAINER ID    IMAGE                           COMMAND                  CREATED      STATUS    PORTS              NAMES
5d1c3f1ca77f    docker.io/library/nginx:latest  "/docker-entrypoint.…"   17
seconds ago     Up       0.0.0.0:80->80/tcp     nginx
```

访问Nginx容器：使用宿主机的IP地址，加上端口号80：

```
[root@containerd ~]# curl http://192.168.1.100:80
<!DOCTYPE html>
<html>
<head>
<title>Welcome to nginx!</title>
<style>
...
</style>
</head>
<body>
<h1>Welcome to nginx!</h1>
...
</body>
</html>
```

查看nerdctl信息：

```
[root@containerd ~]# nerdctl info
Client:
 Namespace:     default
 Debug Mode:    false

Server:
 Server Version: 1.6.33
 Storage Driver: overlayfs
 Logging Driver: json-file
 Cgroup Driver: cgroupfs
 Cgroup Version: 1
 Plugins:
  Log: fluentd journald json-file syslog
  Storage: native overlayfs
 Security Options:
  seccomp
   Profile: builtin
 Kernel Version: 3.10.0-1160.108.1.el7.x86_64
 Operating System: CentOS Linux 7 (Core)
 OSType: linux
 Architecture: x86_64
 CPUs: 1
 Total Memory: 1.794GiB
 Name: containerd
 ID: 384e9c35-60a1-4071-9118-173fed339bc4
```

9.4 使用nerdctl构建镜像

BuildKit是Docker官方推出的一款高效、灵活的构建系统，主要用于构建和打包容器镜像。为了能使用nerdctl构建镜像，需要先安装BuildKit和cni-plugins。

9.4.1 安装 BuildKit 和 cni-plugins

安装BuildKit和cni-plugins的步骤如下：

01 下载BuildKit软件安装包并解压移动：

```
--- 下载软件包
[root@containerd ~]# wget
https://github.com/moby/buildkit/releases/download/v0.14.1/buildkit-v0.14.1.linux-amd64.tar.gz
--- 解压软件包
[root@containerd ~]# tar -xf buildkit-v0.14.1.linux-amd64.tar.gz
--- 将可执行文件移动至/usr/bin目录
[root@containerd ~]# mv bin/* /usr/bin
```

02 创建Buildkit服务，将由Systemd进行管理：

```
[root@containerd ~]# vim /etc/systemd/system/buildkit.service
[Unit]
Description=BuildKit
Documentation=https://github.com/moby/buildkit

[Service]
ExecStart=/usr/bin/buildkitd --oci-worker=false --containerd-worker=true

[Install]
WantedBy=multi-user.target
```

在Linux系统中，特别是在使用Systemd作为系统和服务管理器的系统中，创建/etc/systemd/system/buildkit.service文件是为了定义一个名为buildkit的自定义服务。这样即可使用Systemd来管理服务。buildkit提供了构建镜像的底层功能，但也可以作为独立工具使用。

03 重新加载配置文件并设置开机自启以及查看服务状态：重新加载Systemd的守护进程（daemon）的配置文件，使其能够识别或应用最近对服务（service）单元文件（unit files）所做的更改。

```
[root@containerd ~]# systemctl daemon-reload

--- 设置开机自启并立即启动
[root@containerd ~]# systemctl enable buildkit --now

--- 查看BuildKit的服务状态
[root@containerd ~]# systemctl status buildkit
● buildkit.service - BuildKit
   Loaded: loaded (/etc/systemd/system/buildkit.service; enabled; vendor preset: disabled)
   Active: active (running) since Sun 2024-07-07 21:33:58 CST; 5s ago
     Docs: https://github.com/moby/buildkit
```

```
  Main PID: 12957 (buildkitd)
    Tasks: 7
   Memory: 6.6M
   CGroup: /system.slice/buildkit.service
           └─12957 /usr/bin/buildkitd --oci-worker=false --containerd-worker=true
...
Jul 07 21:33:58 containerd systemd[1]: Started BuildKit.
```

04 下载cni-plugins，创建软件目录并解压该目录：

```
--- 下载软件包
[root@containerd ~]# wget
https://github.com/containernetworking/plugins/releases/download/v1.5.1/cni-plugins-li
nux-amd64-v1.5.1.tgz
--- 创建目录
[root@containerd ~]# mkdir -p /opt/cni/bin
--- 解压至/opt/cni/bin/目录
[root@containerd ~]# tar -xf cni-plugins-linux-amd64-v1.5.1.tgz -C /opt/cni/bin
```

至此，我们完成了BuildKit和cni-plugins的安装。

9.4.2 构建镜像

完成BuildKit和cni-plugins的安装之后，接下来使用nerdctl工具构建镜像。具体步骤如下：

01 创建Dockerfile文件，用于构建一个Nginx镜像：

```
[root@containerd ~]# Dockerfile_Nginx
FROM hub.c.163.com/library/nginx:latest
LABEL author="Dean"
RUN echo "我是院长" >/usr/share/nginx/html/index.html
EXPOSE 80
CMD ["nginx", "-g", "daemon off;"]
```

该Dockerfile基于网易云的镜像仓库（hub.c.163.com）中的library/nginx:latest镜像来构建一个自定义的Nginx镜像。使用LABEL指令设置一个author="Dean"的元数据。然后通过echo输出内容到index.html文件中，并且设置镜像暴露端口为80，最后设置启动Nginx并在前台运行。

02 使用nerdctl工具根据Dockerfile_Nginx文件构建镜像：

```
[root@containerd ~]# nerdctl build -t dean:nginx . -f Dockerfile_Nginx
  [+] Building 0.8s (5/6)                                                    [+]
Building 1.0s (5/6)                                                    [+] Building
1.1s (5/6)                                                    [+] Building 1.3s (5/6)
  [+] Building 1.4s (5/6)
  [+] Building 1.6s (5/6)
  [+] Building 1.7s (5/6)
  [+] Building 1.9s (6/6)
  [+] Building 2.0s (6/6)
   dockerfile: 207B                                             0.0s0
```

```
[+] Building 2.2s (6/6) FINISHED
 => [internal] load build definition from Dockerfile_Nginx    0.0s0
 => => transferring dockerfile:
 => => sending tarball                                        1.2s
```

该命令告诉nerdctl在当前目录下查找名为Dockerfile_Nginx的文件，根据该文件中的指令构建一个新的Docker镜像，并将这个镜像标记为dean:nginx。

03 通过nerdctl工具使用dean:nginx镜像创建容器：

```
[root@containerd ~]# nerdctl run -d -p 80:80 dean:nginx
```

04 访问Nginx容器服务：

```
[root@containerd ~]# curl 127.0.0.1:80
我是院长
```

9.5 本章小结

本章介绍了Containerd的基本概念。首先讲解了如何安装和配置Containerd，然后介绍了如何安装并使用nerdctl工具来对Containerd进行管理。最后介绍了如何安装并使用BuildKit工具来构建镜像。通过本章的学习和实践，读者将能够深入理解和掌握Containerd这个在容器技术领域中占据重要地位的核心组件。

第 10 章 GitLab企业级代码仓库

在软件开发领域，代码管理是项目成功的关键要素之一。随着团队规模的扩大和项目的复杂化，需要一个强大而灵活的代码仓库系统来支持版本控制、团队协作、代码审查、持续集成/持续部署（Continuous Integration/Continuous Deployment，CI/CD）等流程。GitLab作为开源的端到端软件开发生命周期解决方案，凭借其丰富的功能和卓越的性能，成为众多企业首选的代码仓库平台，是运维人员必备的工具之一。本章将介绍GitLab的配置与使用方法。

10.1 GitLab目录结构

GitLab是一个基于Git的版本控制系统，它不仅提供代码仓库功能，还集成了代码管理、持续集成/持续部署（CI/CD）、项目管理、问题跟踪和Wiki等多种功能，形成了一个全方位的开发运维（DevOps）平台。GitLab支持完整的DevOps生命周期，帮助团队实现从代码编写到生产部署的无缝衔接。

下面介绍GitLab的主要目录结构及其关键目录的功能。

（1）库默认存储目录：

`/var/opt/gitlab/git-data/repositories/root`

（2）应用代码和相应的依赖程序：

`/opt/gitlab`

（3）命令编译后的应用数据和配置文件，不需要人为修改配置：

`/var/opt/gitlab`: `gitlab-ctl reconfigure`

（4）配置文件目录：

`/etc/gitlab`

（5）此目录下存放了GitLab各个组件产生的日志：

`/var/log/gitlab`

（6）备份文件生成的目录：

`/var/opt/gitlab/backups/`

GitLab提供了丰富的功能和灵活的目录结构以满足不同的需求。通过了解这些关键目录及其功能，用户可以更好地管理和配置GitLab系统，确保其高效运行。

10.2 部署GitLab

在Kubernetes上部署GitLab，能够充分利用K8s的自动扩展、高可用性和容器化等特性，为GitLab的运行环境提供强大的支撑。通过Kubernetes的声明式配置，可以轻松地管理GitLab的部署、更新和扩展，同时利用K8s的负载均衡和存储卷管理功能，确保GitLab的稳定运行和数据的持久化存储。

本节将通过实例来介绍Kubernetes部署GitLab。由于部署GitLab服务消耗的内存资源较多，我们将升级虚拟机k8s-node1节点内存资源，将内存在原基础上增加4GB，然后部署时将GitLab服务固定到k8s-node1节点运行。

具体操作步骤如下：

01 在k8s-master节点创建资源清单：

```yaml
[root@k8s-master ~]# vim gitlab.yaml
---
apiVersion: v1
kind: Namespace
metadata:
  name: gitlab
---
apiVersion: v1
kind: PersistentVolumeClaim
metadata:
  name: gitlab-postgresql
  namespace: gitlab
spec:
  accessModes:
    - ReadWriteOnce
  resources:
    requests:
      storage: 1Gi
```

```yaml
      storageClassName: dean-nfs   # 指定存储类的名称
---
kind: Service
apiVersion: v1
metadata:
  name: gitlab-postgresql
  namespace: gitlab
  labels:
    name: gitlab-postgresql
spec:
  ports:
    - name: postgres
      protocol: TCP
      port: 5432
      targetPort: postgres
  selector:
    name: postgresql
  type: ClusterIP
---
kind: Deployment
apiVersion: apps/v1
metadata:
  name: gitlab-postgresql
  namespace: gitlab
  labels:
    name: postgresql
spec:
  replicas: 1
  selector:
    matchLabels:
      name: postgresql
  template:
    metadata:
      name: postgresql
      labels:
        name: postgresql
    spec:
      containers:
      - name: postgresql
        image: docker.m.daocloud.io/postgres:12.19
        ports:
        - name: postgres
          containerPort: 5432
        env:
        # 数据库超级用户密码
        - name: POSTGRES_PASSWORD
          value: dean
        # pg数据库用户
        - name: DB_USER
          value: gitlab
```

```yaml
          # pg数据库密码
          - name: DB_PASS
            value: gitlabpwd
          # pg数据库库名
          - name: DB_NAME
            value: gitlab_production
          # 允许所有没有密码的连接
          #- name: POSTGRES_HOST_AUTH_METHOD
          #  value: trust
          - name: DB_EXTENSION
            value: 'pg_trgm,btree_gist'
          # 资源限制
          resources:
            requests:
              cpu: 2
              memory: 1Gi
            limits:
              cpu: 2
              memory: 1Gi
          # 存活探测
          livenessProbe:
            exec:
              command: ["pg_isready","-h","localhost","-U","postgres"]
            initialDelaySeconds: 30
            timeoutSeconds: 5
            periodSeconds: 10
            successThreshold: 1
            failureThreshold: 3
          # 就绪探测
          readinessProbe:
            exec:
              command: ["pg_isready","-h","localhost","-U","postgres"]
            initialDelaySeconds: 5
            timeoutSeconds: 1
            periodSeconds: 10
            successThreshold: 1
            failureThreshold: 3
          # 卷挂载配置
          volumeMounts:
          - name: data
            mountPath: /var/lib/postgresql
      # 卷信息
      volumes:
      - name: data
        persistentVolumeClaim:
          claimName: gitlab-postgresql
---
apiVersion: v1
kind: PersistentVolumeClaim
metadata:
```

```yaml
  name: gitlab-redis
  namespace: gitlab
spec:
  accessModes:
    - ReadWriteOnce
  resources:
    requests:
      storage: 1Gi
  storageClassName: dean-nfs  # 指定存储类的名称
---
kind: Service
apiVersion: v1
metadata:
  name: gitlab-redis
  namespace: gitlab
  labels:
    name: gitlab-redis
spec:
  type: ClusterIP
  ports:
    - name: redis
      protocol: TCP
      port: 6379
      targetPort: redis
  selector:
    name: gitlab-redis
---
kind: Deployment
apiVersion: apps/v1
metadata:
  name: gitlab-redis
  namespace: gitlab
  labels:
    name: gitlab-redis
spec:
  replicas: 1
  selector:
    matchLabels:
      name: gitlab-redis
  template:
    metadata:
      name: gitlab-redis
      labels:
        name: gitlab-redis
    spec:
      containers:
      - name: gitlab-redis
        image: docker.m.daocloud.io/redis:7.2.5
        imagePullPolicy: IfNotPresent
        ports:
```

```yaml
        - name: redis
          containerPort: 6379
          protocol: TCP
        resources:
          limits:
            cpu: 1000m
            memory: 1Gi
          requests:
            cpu: 1000m
            memory: 1Gi
        volumeMounts:
          - name: data
            mountPath: /var/lib/redis
        livenessProbe:
          exec:
            command:
              - redis-cli
              - ping
          initialDelaySeconds: 5
          timeoutSeconds: 5
          periodSeconds: 10
          successThreshold: 1
          failureThreshold: 3
        readinessProbe:
          exec:
            command:
              - redis-cli
              - ping
          initialDelaySeconds: 5
          timeoutSeconds: 5
          periodSeconds: 10
          successThreshold: 1
          failureThreshold: 3
      volumes:
        - name: data
          persistentVolumeClaim:
            claimName: gitlab-redis
---
apiVersion: v1
kind: PersistentVolumeClaim
metadata:
  name: gitlab-config
  namespace: gitlab
spec:
  accessModes:
    - ReadWriteOnce
  resources:
    requests:
      storage: 1Gi
  storageClassName: dean-nfs    # 指定存储类的名称
```

```yaml
---
apiVersion: v1
kind: PersistentVolumeClaim
metadata:
  name: gitlab-logs
  namespace: gitlab
spec:
  accessModes:
    - ReadWriteOnce
  resources:
    requests:
      storage: 1Gi
  storageClassName: dean-nfs    # 指定存储类的名称
---
apiVersion: v1
kind: PersistentVolumeClaim
metadata:
  name: gitlab-data
  namespace: gitlab
spec:
  accessModes:
    - ReadWriteOnce
  resources:
    requests:
      storage: 1Gi
  storageClassName: dean-nfs    # 指定存储类的名称
---
kind: Service
apiVersion: v1
metadata:
  name: gitlab
  namespace: gitlab
  labels:
    name: gitlab
spec:
  ports:
    - name: http
      protocol: TCP
      port: 80
      targetPort: http
      nodePort: 31080
    - name: ssh
      protocol: TCP
      port: 22
      targetPort: ssh
      nodePort: 31022
  selector:
    name: gitlab
  type: NodePort
---
```

```yaml
kind: Deployment
apiVersion: apps/v1
metadata:
  name: gitlab
  namespace: gitlab
  labels:
    name: gitlab
spec:
  replicas: 1
  selector:
    matchLabels:
      name: gitlab
  template:
    metadata:
      name: gitlab
      labels:
        name: gitlab
    spec:
      # 为了顺利部署测试，增加了k8s-node1节点内存，所以将Pod调度到k8s-node1节点
      nodeName: k8s-node1
      containers:
      - name: gitlab
        image: gitlab/gitlab-ce:15.6.8-ce.0
        imagePullPolicy: IfNotPresent
        ports:
        - name: ssh
          containerPort: 22
        - name: http
          containerPort: 80
        - name: https
          containerPort: 443
        # 环境变量
        env:
        - name: TZ
          value: Asia/Shanghai
        - name: GITLAB_TIMEZONE
          value: Beijing
        - name: GITLAB_SECRETS_DB_KEY_BASE
          value: long-and-random-alpha-numeric-string
        - name: GITLAB_SECRETS_SECRET_KEY_BASE
          value: long-and-random-alpha-numeric-string
        - name: GITLAB_SECRETS_OTP_KEY_BASE
          value: long-and-random-alpha-numeric-string
        # GitLab的Web密码
        - name: GITLAB_ROOT_PASSWORD
          value: deanhandsome
        - name: GITLAB_ROOT_EMAIL
          value: deanmr@qq.com
        - name: GITLAB_HOST
          value: 'gitlab.deanit.cn'
```

```yaml
        - name: GITLAB_PORT
          value: '80'
        - name: GITLAB_SSH_PORT
          value: '22'
        - name: GITLAB_NOTIFY_ON_BROKEN_BUILDS
          value: 'true'
        - name: GITLAB_NOTIFY_PUSHER
          value: 'false'
        - name: GITLAB_BACKUP_TIME
          value: 01:00
        - name: DB_TYPE
          value: postgres
        - name: DB_HOST
          value: gitlab-postgresql
        - name: DB_PORT
          value: '5432'
        - name: DB_USER
          value: gitlab
        - name: DB_PASS
          value: gitlabpwd
        - name: DB_NAME
          value: gitlab_production
        - name: REDIS_HOST
          value: gitlab-redis
        - name: REDIS_PORT
          value: '6379'
        # 资源限制
        #resources:
        #  limits:
        #    cpu: 2
        #    memory: 4Gi
        #  requests:
        #    cpu: 2
        #    memory: 4Gi
        # 探针
        #livenessProbe:
        #  httpGet:
        #    path: /
        #    port: 80
        #    scheme: HTTP
        #  initialDelaySeconds: 300
        #  timeoutSeconds: 5
        #  periodSeconds: 10
        #  successThreshold: 1
        #  failureThreshold: 3
        #readinessProbe:
        #  httpGet:
        #    path: /
        #    port: 80
        #    scheme: HTTP
```

```yaml
#   initialDelaySeconds: 5
#   timeoutSeconds: 30
#   periodSeconds: 10
#   successThreshold: 1
#   failureThreshold: 3
        volumeMounts:
        - name: gitlab-config
          mountPath: /etc/gitlab
        - name: gitlab-logs
          mountPath: /var/log/gitlab
        - name: gitlab-data
          mountPath: /var/opt/gitlab
        - name: localtime
          mountPath: /etc/localtime
      volumes:
      - name: gitlab-config
        persistentVolumeClaim:
          claimName: gitlab-config
      - name: gitlab-logs
        persistentVolumeClaim:
          claimName: gitlab-logs
      - name: gitlab-data
        persistentVolumeClaim:
          claimName: gitlab-data
      - name: localtime
        hostPath:
          path: /etc/localtime
---
apiVersion: networking.k8s.io/v1
kind: Ingress
metadata:
  name: gitlab
  namespace: gitlab
spec:
  ingressClassName: nginx
  rules:
  - host: gitlab.deanit.cn
    http:
      paths:
      - backend:
          service:
            name: gitlab
            port:
              number: 80
        path: /
        pathType: Prefix
```

02 应用资源：

```
[root@k8s-master ~]# kubectl apply -f gitlab.yaml
namespace/gitlab created
```

```
persistentvolumeclaim/gitlab-postgresql created
service/gitlab-postgresql created
deployment.apps/gitlab-postgresql created
persistentvolumeclaim/gitlab-redis created
service/gitlab-redis created
deployment.apps/gitlab-redis created
persistentvolumeclaim/gitlab-config created
persistentvolumeclaim/gitlab-logs created
persistentvolumeclaim/gitlab-data created
service/gitlab created
deployment.apps/gitlab created
ingress.networking.k8s.io/gitlab created
```

03 这里使用Kubectl工具将gitlab.yaml文件中定义的资源应用到Kubernetes集群中。使用-f选项指定要应用的文件名。

根据输出信息，可以看出已经成功创建了以下资源：

```
namespace/gitlab created
persistentvolumeclaim/gitlab-postgresql created
service/gitlab-postgresql created
deployment.apps/gitlab-postgresql created
persistentvolumeclaim/gitlab-redis created
service/gitlab-redis created
deployment.apps/gitlab-redis created
persistentvolumeclaim/gitlab-config created
persistentvolumeclaim/gitlab-logs created
persistentvolumeclaim/gitlab-data created
service/gitlab created
deployment.apps/gitlab created
ingress.networking.k8s.io/gitlab created
```

04 查看GitLab服务各资源的状态，如图10-1所示。

图 10-1　GitLab 服务各资源的状态

05 由于我们创建了ingress资源，可以通过域名来访问GitLab服务，各客户端配置如下：

```
--- Linux和macOS客户端配置
$ vim /etc/hosts
192.168.1.11 gitlab.deanit.cn

--- Windows客户端配置
# hosts文件位于路径: C:\Windows\System32\drivers\etc\hosts
192.168.1.11 gitlab.deanit.cn
```

> **注意** 由于hosts文件属于系统文件，因此在修改之前需要获取管理员权限。

/etc/hosts文件是Linux和UNIX系统中的一个非常重要的配置文件，它用于将主机名映射到相应的IP地址。这个文件通常用于解决网络中的域名和IP地址之间的对应关系，使得系统可以基于主机名来访问网络上的资源，而不需要直接记住或输入IP地址。

06 通过域名访问后，效果如图10-2所示。

图10-2　访问 GitLab 首页的效果

07 账号和密码是在YAML文件中定义的，账号为root，密码为deanhandsome。登录成功后如图10-3所示。

图10-3　成功登录 GitLab 首页

10.3　GitLab的配置与使用

本节将介绍GitLab的配置和使用方法。

10.3.1　基础设置

01　登录成功后，默认是英文页面，将其改为简体中文。单击右上角的用户头像，再单击Preferences，如图10-4所示。

图10-4　单击 Preferences 按钮

02　找到Language选项，将其改为简体中文，然后单击Save changes按钮进行保存，如图10-5所示。

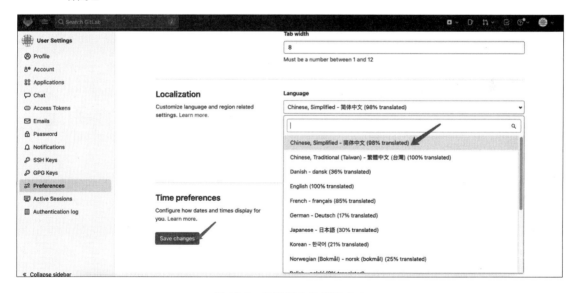

图10-5　设置语言为简体中文

03　刷新页面，即可显示简体中文，官方标识了98%已翻译，可能存在有些地方还是英文，如图10-6所示。

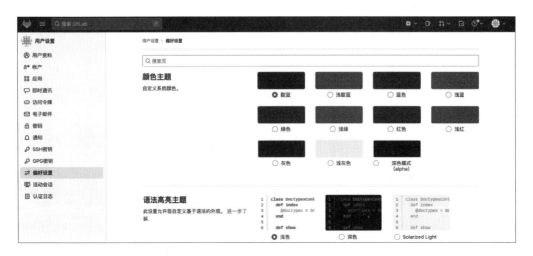

图 10-6　页面已显示为简体中文

10.3.2　创建项目

01　通过首页新建项目，如图10-7所示。

图 10-7　新建项目

02　单击"创建空白项目"按钮，如图10-8所示。

图 10-8　创建空白项目

03 输入项目名称和项目标识串,然后单击"新建项目"按钮,如图10-9所示。

图 10-9　设置项目信息

04 项目创建后如图10-10所示。

图 10-10　项目创建成功后的效果

10.3.3　修改克隆地址

01 当克隆项目时发现地址是默认的,并不是我们配置的网址,就需要修改配置,如图10-11所示。

02 查看GitLab配置的PV名字:

```
[root@k8s-master nfs]# kubectl get pv | grep Bound | grep gitlab-config | awk '{print $1}'
```

```
# 输出信息
pvc-c1fedcd8-48fe-4015-badd-1a791bbc164d
```

图 10-11　GitLab 拉取地址信息

03 使用Kubectl命令行工具来查询Kubernetes集群中状态为Bound的PersistentVolume（PV）资源，并特别筛选出那些名称中包含gitlab-config的PV，最后只打印出这些PV的名称。

04 之前规划的k8s-master节点作为NFS服务端，存储目录为/nfs，找到该存储卷目录：

```
--- 进入GitLab的配置文件存储目录
[root@k8s-master ~]# cd
/nfs/gitlab-gitlab-config-pvc-c1fedcd8-48fe-4015-badd-1a791bbc164d/

--- 查看配置文件目录下的文件
[root@k8s-master gitlab-gitlab-config-pvc-c1fedcd8-48fe-4015-badd-1a791bbc164d]# ll
总用量 184
-rw------- 1 root root 139778 7月  13 16:20 gitlab.rb
-rw------- 1 root root  19349 7月  13 16:43 gitlab-secrets.json
-rw------- 1 root root    525 7月  13 16:20 ssh_host_ecdsa_key
-rw-r--r-- 1 root root    190 7月  13 16:20 ssh_host_ecdsa_key.pub
-rw------- 1 root root    419 7月  13 16:20 ssh_host_ed25519_key
-rw-r--r-- 1 root root    110 7月  13 16:20 ssh_host_ed25519_key.pub
-rw------- 1 root root   2622 7月  13 16:20 ssh_host_rsa_key
-rw-r--r-- 1 root root    582 7月  13 16:20 ssh_host_rsa_key.pub
drwxr-xr-x 2 root root      6 7月  13 16:20 trusted-certs
```

05 编辑gitlab.rb文件，修改external_url字段信息：将#external_url 'GENERATED_EXTERNAL_URL'取消注释，并修改为：external_url 'http://gitlab.deanit.cn'。

> **注意** http://必须要写，不然重启Pod会报错。

在GitLab中，external_url是一个重要的配置项，它定义了GitLab实例从外部网络访问的URL。这个设置对于确保GitLab能够正确处理重定向、构建正确的链接（如项目克隆链接）以及保证OAuth等外部认证机制的正常工作至关重要。

06 删除旧Pod使之配置生效：

```
--- 查看Pod名
[root@k8s-master ~]# kubectl get pods -n gitlab
NAME                                    READY   STATUS    RESTARTS      AGE
gitlab-57b89f6f88-mbvxk                 1/1     Running   1 (45m ago)   68m
gitlab-postgresql-794db56bf6-c2zb7      1/1     Running   0             68m
gitlab-redis-85bd6bdb6-75f2f            1/1     Running   0             68m

--- 删除Pod
[root@k8s-master ~]# kubectl delete pods -n gitlab gitlab-57b89f6f88-mbvxk
pod "gitlab-57b89f6f88-mbvxk" deleted

--- 再次查看Pod状态，已经启动完成
[root@k8s-master ~]# kubectl get pods -n gitlab
NAME                                    READY   STATUS    RESTARTS   AGE
gitlab-57b89f6f88-v88hl                 1/1     Running   0          2m
gitlab-postgresql-794db56bf6-c2zb7      1/1     Running   0          71m
gitlab-redis-85bd6bdb6-75f2f            1/1     Running   0          71m
```

07 再次刷新网页，即可看到克隆地址已经改为网址，如图10-12所示。

图10-12　GitLab 正确地拉取地址信息

10.3.4 拉取/提交代码

在客户端通过git命令工具克隆项目进行测试：

```
--- 克隆项目
dean@handsome:~|
⇒  git clone http://gitlab.deanit.cn/root/deanitweb.git
Cloning into 'deanitweb'...
Username for 'http://gitlab.deanit.cn': root
Password for 'http://root@gitlab.deanit.cn':
remote: Enumerating objects: 3, done.
remote: Counting objects: 100% (3/3), done.
remote: Compressing objects: 100% (2/2), done.
remote: Total 3 (delta 0), reused 0 (delta 0), pack-reused 0
Receiving objects: 100% (3/3), done.

--- 进入项目
dean@handsome:~|
⇒  cd deanitweb

--- 创建一个index.html文件并写入内容i am dean
dean@handsome:~/deanitweb|
⇒  echo "i am dean">index.html

--- 提交至暂存区
dean@handsome:~/deanitweb|
⇒  git add .

--- 查看提交状态
dean@handsome:~/deanitweb|
⇒  git status
On branch main
Your branch is up to date with 'origin/main'.

Changes to be committed:
  (use "git restore --staged <file>..." to unstage)
    new file:   index.html

--- 将暂存区中的改动内容提交到仓库中并提交说明信息
dean@handsome:~/deanitweb|
⇒  git commit -m "第一次提交代码"

[main 242c8b8] 第一次提交代码
 1 file changed, 1 insertion(+)
 create mode 100644 index.html

--- 进行推送代码
dean@handsome:~/deanitweb|
⇒  git push
Enumerating objects: 4, done.
```

```
Counting objects: 100% (4/4), done.
Delta compression using up to 8 threads
Compressing objects: 100% (2/2), done.
Writing objects: 100% (3/3), 301 bytes | 301.00 KiB/s, done.
Total 3 (delta 0), reused 0 (delta 0), pack-reused 0
To http://gitlab.deanit.cn/root/deanitweb.git
   d9fcb87..242c8b8  main -> main
```

参数解析：

- git clone [url]：克隆一个远程仓库到本地。
- git add [file]：将指定的文件添加到暂存区，准备提交。使用git add .可以添加当前目录下的所有修改和新增的文件。
- git commit -m "message"：提交暂存区的文件到本地仓库，并附上一条描述本次提交的备注信息。
- git status：显示工作区和暂存区的状态，包括哪些文件已被修改但还未提交等。
- git push [remote] [branch]：将本地分支推送到远程仓库。

再次刷新网页查看项目信息，可以看到我们提交的代码信息，如图10-13所示。

图10-13　代码提交信息和提交内容

10.4 本章小结

本章详细介绍了如何在Kubernetes集群上部署GitLab这一流行的开源Git仓库管理工具。GitLab不仅提供了Git仓库的托管服务，还集成了代码审查、CI/CD、项目管理等多种功能，是现代软件开发团队不可或缺的工具之一。接下来，介绍了如何使用Git工具通过命令行操作GitLab来提交克隆和提交代码。通过本章的学习，读者将能够掌握在Kubernetes上部署GitLab的基本流程和关键步骤，为在云原生环境中构建高效、可扩展的软件开发环境打下坚实的基础。

第 11 章 Jenkins持续集成交付工具

Jenkins作为一个开源的自动化服务器，已经成为现代软件开发和持续集成/持续部署（CI/CD）流程中不可或缺的一部分，以其强大的功能、灵活的配置选项以及丰富的插件生态系统，赢得了全球范围内开发者和运维团队的青睐。本章将全面概述Jenkins的基本概念、核心功能、应用场景、架构原理及其在软件开发流程中的作用。

11.1 Jenkins概述

Jenkins是一款基于Java语言开发的开源软件工具，专门用于自动化处理软件开发过程中的多项任务。这些任务涵盖构建、测试、打包、部署以及监控等关键环节。通过其用户友好的Web界面和RESTful API接口，Jenkins极大地简化了自动化流程的配置与管理过程。更值得一提的是，Jenkins还引入了分布式构建机制，允许在多个计算节点上并行执行任务，从而显著提升了构建效率和资源利用效率。

Jenkins的核心功能如下。

- 自动化构建：Jenkins能够实时监测代码仓库（如Git、SVN）的变化，一旦检测到新代码提交，便会自动触发构建流程。
- 持续集成：借助自动化的构建与测试流程，Jenkins助力开发团队实现持续集成，确保新增代码不会破坏现有功能，保障软件的稳定性与可靠性。
- 丰富的插件生态：Jenkins拥有一个庞大的插件库，覆盖从代码质量分析到云部署等各个方面的需求，使用户能够根据具体需求定制个性化的自动化流程。
- 分布式构建支持：Jenkins支持分布式构建，能够在多个节点上并行执行任务，这不仅加快了构建速度，还提高了资源的使用效率。

- 可视化报告：Jenkins提供详尽的构建报告和测试结果，帮助用户迅速定位问题并优化代码质量，提升开发效率。

11.2 Kubernetes集群部署Jenkins

本节介绍如何在Kubernetes集群中部署Jenkins，具体步骤如下：

01 在k8s-master节点创建资源清单：

```
[root@k8s-master ~]# vim jenkins.yaml
---
# 命名空间
apiVersion: v1
kind: Namespace
metadata:
  name: jenkins
---
# 授权
apiVersion: rbac.authorization.k8s.io/v1
kind: ClusterRoleBinding
metadata:
  name: jenkins
  namespace: jenkins
roleRef:
  apiGroup: rbac.authorization.k8s.io
  kind: ClusterRole
  name: cluster-admin
subjects:
- kind: ServiceAccount
  name: default
  namespace: jenkins
---
# pvc
apiVersion: v1
kind: PersistentVolumeClaim
metadata:
  name: jenkins
  namespace: jenkins
spec:
  storageClassName: "dean-nfs"
  accessModes:
    - ReadWriteMany
  resources:
    requests:
      storage: 5Gi
---
# 控制器
apiVersion: apps/v1
```

```yaml
kind: Deployment
metadata:
  name: jenkins
  namespace: jenkins
spec:
  replicas: 1
  selector:
    matchLabels:
      app: jenkins
  template:
    metadata:
      labels:
        app: jenkins
    spec:
      nodeName: k8s-node1
      containers:
      - name: jenkins
        image: docker.m.daocloud.io/jenkins/jenkins:2.452.3-lts-jdk11
        imagePullPolicy: IfNotPresent
        ports:
        - containerPort: 8080
          name: web
          protocol: TCP
        - containerPort: 50000
          name: agent
          protocol: TCP
        resources:
          limits:
            cpu: 1000m
            memory: 1Gi
          requests:
            cpu: 1000m
            memory: 1Gi
        livenessProbe:                     # 存活探测
          httpGet:
            #path: /dean/login             # 设置了--prefix=/dean参数，就需要改为探测接口
            path: /login
            port: 8080
          # 容器初始化完成后，等待240秒进行探针检查，根据实际情况进行配置
          initialDelaySeconds: 240
          timeoutSeconds: 5                # 超时时间
          failureThreshold: 12             # 当Pod成功启动且检查失败时，K8s将在放弃之前尝试
failureThreshold次。放弃生存检查意味着重新启动Pod。而放弃就绪检查，Pod将被标记为未就绪。默认为3，最小值为1
        # 就绪探测
        readinessProbe:
          httpGet:
            #path: /dean/login
            path: /login
            port: 8080
```

```yaml
        initialDelaySeconds: 240
        timeoutSeconds: 5
        failureThreshold: 12
      volumeMounts:
      - name: jenkins-home
        mountPath: /var/jenkins_home
      env:
      - name: LIMITS_MEMORY
        valueFrom:
          resourceFieldRef:
            resource: limits.memory
            divisor: 1Mi
      # Jenkins参数设置，根据自身情况是否需要配置
      # - name: "JENKINS_OPTS"
      # 设置路径前缀加上dean,访问时如http://ip/dean
      #   value: "--prefix=/dean"
      # JVM 参数设置，设置变量，指定时区
      - name: JAVA_OPTS
        value: "
                -Xmx$(LIMITS_MEMORY)m
                -XshowSettings:vm
                -Dhudson.slaves.NodeProvisioner.initialDelay=0
                -Dhudson.slaves.NodeProvisioner.MARGIN=50
                -Dhudson.slaves.NodeProvisioner.MARGIN0=0.85
                -Duser.timezone=Asia/Shanghai
                -Djava.awt.headless=true
                -Dhudson.security.csrf.GlobalCrumbIssuerConfiguration.DISABLE_CSRF_PROTECTION=true
                "
      securityContext:
        fsGroup: 1000
        runAsUser: 1000          # 设置以ROOT用户运行容器
        #privileged: true        # 拥有特权
      volumes:
      - name: jenkins-home
        persistentVolumeClaim:
          claimName: jenkins
---
# svc服务
apiVersion: v1
kind: Service
metadata:
  name: jenkins
  namespace: jenkins
  labels:
    app: jenkins
spec:
  selector:
    app: jenkins
  type: NodePort
```

```yaml
  ports:
  - name: web
    port: 8080
    targetPort: web
    nodePort: 30012
  - name: agent
    port: 50000
    targetPort: agent
---
apiVersion: networking.k8s.io/v1
kind: Ingress
metadata:
  name: jenkins
  namespace: jenkins
spec:
  ingressClassName: nginx
  rules:
  - host: jenkins.deanit.cn
    http:
      paths:
      - backend:
          service:
            name: jenkins
            port:
              number: 8080
        path: /
        pathType: Prefix
```

02 应用资源：

```
[root@k8s-master ~]# kubectl apply -f jenkins.yaml
namespace/jenkins created
clusterrolebinding.rbac.authorization.k8s.io/jenkins created
persistentvolumeclaim/jenkins created
deployment.apps/jenkins created
service/jenkins created
ingress.networking.k8s.io/jenkins created
```

03 使用Kubectl工具将jenkins.yaml文件中定义的资源应用到Kubernetes集群中。

根据输出信息，可以看出已经成功创建了以下资源：

```
namespace/jenkins created
clusterrolebinding.rbac.authorization.k8s.io/jenkins created
persistentvolumeclaim/jenkins created
deployment.apps/jenkins created
service/jenkins created
ingress.networking.k8s.io/jenkins created
```

04 查看Jenkins服务各资源的状态，如图11-1所示。

```
[root@k8s-master ~]# kubectl get pods,svc,ing,pvc,pv -n jenkins
NAME                          READY   STATUS    RESTARTS   AGE
pod/jenkins-68ffd8787c-j5mwt  1/1     Running   0          11m

NAME             TYPE       CLUSTER-IP    EXTERNAL-IP   PORT(S)                         AGE
service/jenkins  NodePort   10.98.78.125  <none>        8080:30012/TCP,50000:32875/TCP  11m

NAME                             CLASS   HOSTS               ADDRESS         PORTS   AGE
ingress.networking.k8s.io/jenkins nginx  jenkins.deanit.cn   10.110.206.243  80      2m9s

NAME                              STATUS   VOLUME                                    CAPACITY   ACCESS MODES   STORAGECLASS   AGE
persistentvolumeclaim/jenkins     Bound    pvc-6df745f4-21f7-4188-af1c-9259fb0fa371   5Gi        RWX            dean-nfs       11m

NAME                                                         CAPACITY   ACCESS MODES   RECLAIM POLICY   STATUS   CLAIM             STORAGECLASS   REASON   AGE
persistentvolume/pvc-6df745f4-21f7-4188-af1c-9259fb0fa371    5Gi        RWX            Retain           Bound    jenkins/jenkins   dean-nfs                11m
```

图 11-1　Jenkins 服务各资源的状态

05 查看Jenkins初始密码：

```
[root@k8s-master ~]# kubectl exec -it -n jenkins `kubectl get pods -n jenkins | grep jenkins | awk '{print $1}'` cat /var/jenkins_home/secrets/initialAdminPassword
# 输出信息
2348e011a0fe468b8b131ac41d01d2bc
```

该命令是在jenkins命名空间中，找到所有包含jenkins字符串的Pod，并取第一个Pod的名称，然后在这个Pod中执行cat命令来查看/var/jenkins_home/secrets/initialAdminPassword文件的内容，即Jenkins的初始管理员密码。

06 由于我们创建了ingress资源，因此可以通过域名来访问Jenkins服务，各客户端配置如下：

```
# Linux和macOS客户端配置
$ vim /etc/hosts
192.168.1.11 jenkins.deanit.cn


# Windows客户端配置
# hosts文件位于路径：C:\Windows\System32\drivers\etc\hosts。
192.168.1.11 jenkins.deanit.cn
```

运维笔记：由于hosts文件属于系统文件，因此在修改之前需要获取管理员权限。

07 通过域名访问Jenkins，输入上面获取的Jenkins初始密码单击"继续"按钮，如图11-2所示。

图 11-2　Jenkins 初始化页面

08 这里单击"选择插件来安装"按钮，如图11-3所示。

第 11 章　Jenkins 持续集成交付工具　197

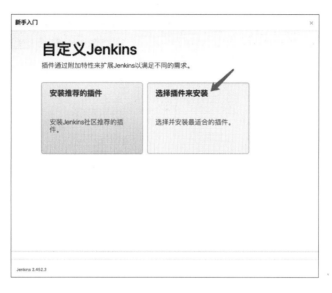

图 11-3　自定义 Jenkins 插件选择页面

09　选择语言插件，单击"安装"按钮，如图 11-4 所示。

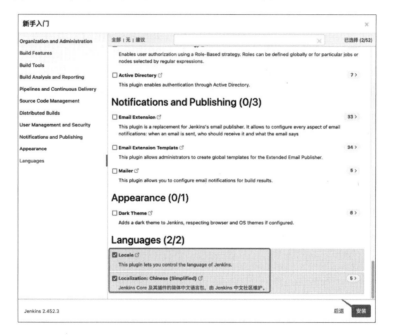

图 11-4　Jenkins 插件安装页面

10　插件安装中，如图 11-5 所示。
11　上面的插件安装完成后会跳转到如图 11-6 所示的页面，输入我们要创建的用户信息，单击"保存并完成"按钮。

图 11-5　Jenkins 插件安装中

图 11-6　Jenkins 创建管理员用户

⑫ Jenkins实例配置，保持默认设置，单击"保存并完成"按钮，如图11-7所示。

⑬ 单击"开始使用Jenkins"按钮，如图11-8所示。

图 11-7　Jenkins 实例配置

图 11-8　Jenkins 已就绪

⑭ 进入Jenkins首页，如图11-9所示。

至此，我们已经成功在Kubernetes集群中部署了Jenkins。

第 11 章　Jenkins 持续集成交付工具　199

图 11-9　Jenkins 首页

11.3　Jenkins对接K8s实现动态Slave

Jenkins作为业界领先的自动化服务器，通过与Kubernetes的深度集成，能够实现动态Slave（即动态代理节点），从而提供更加灵活和高效的构建、测试及部署能力。本节将详细探讨如何配置Jenkins以对接K8s，并分别通过自由风格项目和Pipeline流水线项目来实现动态Slave，旨在帮助读者掌握这一关键技术，优化CI/CD流程，提升开发效率。

11.3.1　基础设置并对接 K8s

01　依次打开Manage Jenkins→"插件管理"，搜索kubernetes，勾选kubernetes插件并单击"安装"按钮，如图11-10所示。

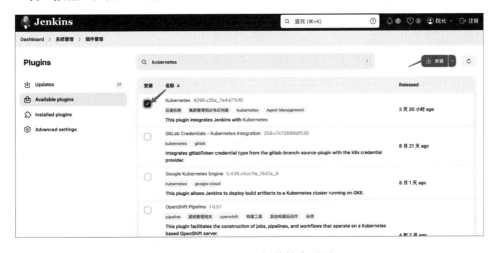

图 11-10　Jenkins 插件搜索页面

02 安装完成后重启Jenkins（空闲时），如图11-11所示。

图11-11　Jenkins重启页面

03 因为配置的就绪探测和存活探测为4分钟，所以重启完成时间为4分钟，如图11-12所示。

图11-12　Jenkins登录页面

04 依次单击"系统管理"→"节点和云管理"→Clouds→New cloud进行新建，如图11-13所示。

图11-13　节点列表页面

05 输入添加的云名称，类型选择Kubernetes，然后单击Create按钮，如图11-14所示。

06 输入云名称、Kubernetes地址、Kubernetes命名空间以及Jenkins地址，单击"连接测试"按钮，如图11-15所示。

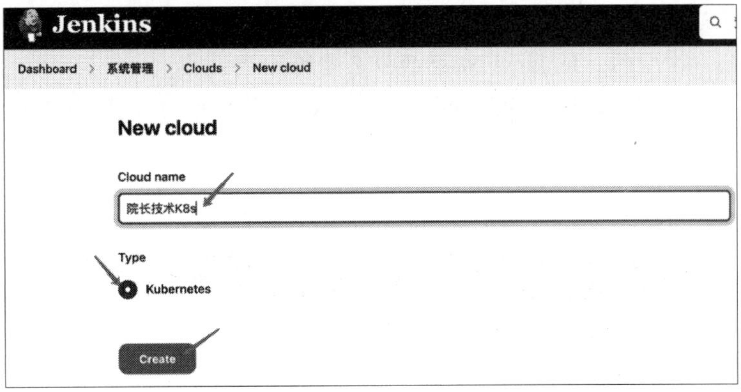

图 11-14 新建云节点名称和类型

图 11-15 配置云节点信息

参数解析：

- 名称：添加的云名称。

- Kubernetes地址：https://kubernetes.default.svc.cluster.local，短域名：https://kubernetes.default。
- Kubernetes命名空间：Jenkins所部署的命名空间。
- Jenkins地址：http://jenkins.jenkins.svc.cluster.local:8080，格式为：服务名.命名空间.svc.cluster.local:8080。

07 单击Save按钮保存后，接下来配置Pod模板，如图11-16所示。

图 11-16　配置 Pod 模板信息

08 配置容器模板，容器镜像中的Java版本一定要和Jenkins的Java版本保持一致，如图11-17所示。

图 11-17　配置容器模板信息

09 配置Pod挂载卷信息，如图11-18所示。

图 11-18　配置 Pod 挂载卷信息

11.3.2　自由风格项目实现动态 Slave

01 在Jenkins首页新建任务，输入任务名称，选择构建一个自由风格的软件项目，如图11-19所示。

图 11-19　新建任务

02 输入标签列表：jenkins-agent，如图11-20所示。

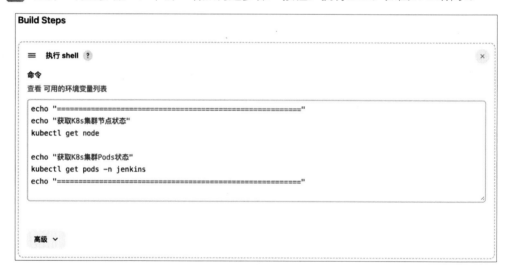

图 11-20　配置限制项目的运行节点

03 选择"构建步骤"，单击"增加构建步骤"按钮，执行shell，如图11-21所示。

图 11-21　执行 shell 命令配置

04 保存后，单击左侧的"立即构建"按钮开始构建过程，控制台输出结果如图11-22所示。

05 在构建任务时，会创建一个名为jenkins-agent-xxxxx的Pod，执行完任务会自动删除：

```
[root@k8s-master ~]# kubectl get pods -n jenkins
NAME                        READY   STATUS    RESTARTS   AGE
jenkins-5d89b6c8fd-8wcrb    1/1     Running   0          7h25m
jenkins-agent-53hf3         1/1     Running   0          12s
```

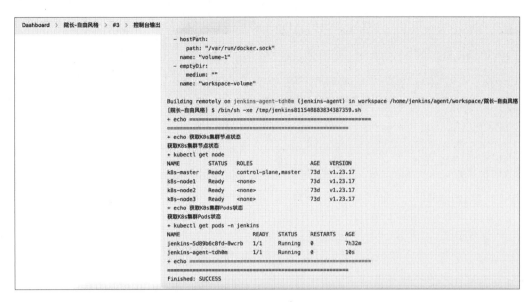

图 11-22　控制台输出信息

11.3.3　Pipeline 流水线项目实现动态 Slave

Pipeline的实现方式是一套Groovy DSL，任何发布流程都可以表述为一段Groovy脚本，并且Jenkins支持从代码库直接读取脚本，从而实现了Pipeline as Code的理念。

01 在首页新建任务，输入任务名称，选择流水线，如图11-23所示。

图 11-23　新建任务

02 在流水线步骤中,选择定义类型为Pipeline script,并输入以下Pipeline脚本。

```
pipeline{
    agent{
        kubernetes{
            label "jenkins-agent"
            yaml '''
---
kind: Pod
apiVersion: v1
metadata:
  labels:
    label: jenkins-agent
  name: jenkinsagent
  namespace: jenkins
spec:
containers:
  - name: jenkinsagent
    image: docker.m.daocloud.io/jenkins/inbound-agent:bookworm-jdk11
    imagePullPolicy: IfNotPresent
    resources:
      limits:
        cpu: 1000m
        memory: 2Gi
      requests:
        cpu: 500m
        memory: 512Mi
'''
        }
    }

    stages{
        stage("test"){
            steps{
                script{
                    sh "java -version & sleep 30"
                }
            }
        }
    }
}
```

添加Pipeline内容后如图11-24所示。

03 保存后,单击左侧的"立即构建"按钮开始构建过程,控制台输出结果如图11-25所示。

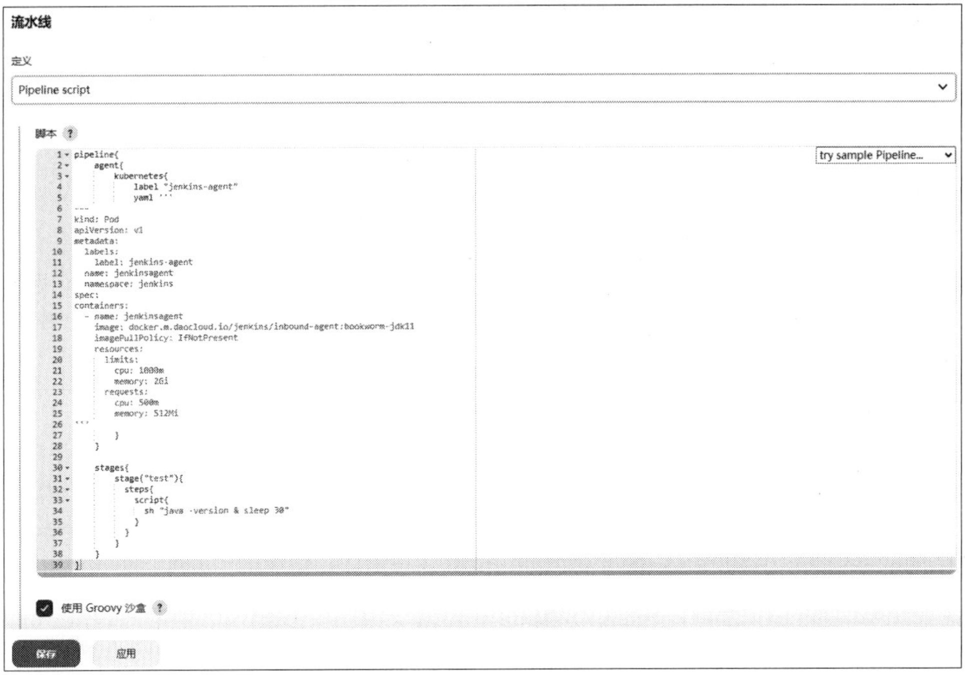

图 11-24 Pipeline 脚本

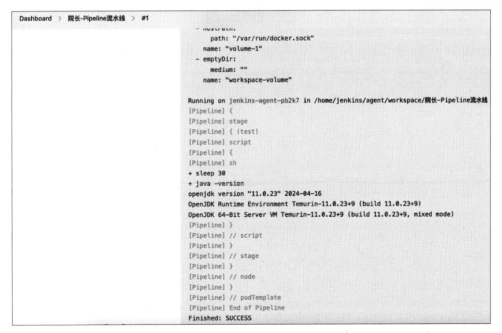

图 11-25 控制台输出信息

04 在构建任务时，会创建一个名为jenkins-agent-xxxxx的Pod，执行完任务会自动删除：

```
[root@k8s-master ~]# kubectl get pods -n jenkins
NAME                        READY   STATUS    RESTARTS   AGE
jenkins-5d89b6c8fd-8wcrb    1/1     Running   0          7h46m
jenkins-agent-pb2k7         1/1     Running   0          15s
```

11.4 本章小结

本章介绍了如何在Kubernetes环境中安装Jenkins，并详细阐述了如何将Jenkins与Kubernetes集群对接以实现动态Slave（工作节点）的创建与管理。首先，介绍了Jenkins概述以及核心功能。随后，深入探讨了如何在Kubernetes集群上部署Jenkins，这包括直接通过Kubernetes的YAML配置文件手动部署Jenkins服务。在Jenkins成功部署到Kubernetes之后，重点讲解了如何配置Jenkins以使其能够动态地利用Kubernetes集群中的资源来创建Slave节点。这一过程涉及Jenkins的Kubernetes插件配置，该插件允许Jenkins根据作业需求自动在Kubernetes集群中启动和销毁Pod作为slave节点，从而实现了资源的按需分配和高效利用。我们详细说明了如何配置Jenkins以识别Kubernetes集群，以及定义slave Pod的模板。通过本章的学习，读者将能够掌握在Kubernetes环境中安装和配置Jenkins的完整流程，以及如何利用Jenkins与Kubernetes的集成来实现动态slave的创建与管理，从而优化CI/CD流程，提高软件开发和部署的效率。

第 12 章 ArgoCD声明式持续交付

随着应用程序的复杂度和规模不断增长,如何高效地管理这些在Kubernetes上运行的应用,确保它们的稳定性、可靠性和快速迭代,成为开发者和运维团队面临的重大挑战。ArgoCD作为一款强大的开源工具,通过声明式的方式实现了持续交付,使得Kubernetes集群的自动化部署变得更加简单和高效。

本章将深入探讨ArgoCD的核心特性、安装配置以及实际使用场景,帮助读者掌握如何利用ArgoCD来优化他们的Kubernetes工作流程。

12.1 ArgoCD概述

ArgoCD作为一个基于GitOps声明的持续交付(Continuous Delivery,CD)工具,为Kubernetes应用程序的自动化部署和管理提供了强大的支持。通过声明式配置管理、自动化部署与同步、健康监测与回滚等核心特性,ArgoCD能够显著提高软件交付的效率和质量。同时,其可扩展性和定制化特性也使其能够适应不同规模和复杂度的应用场景。

ArgoCD和Jenkins在实现持续集成和持续部署(CI/CD)方面各有其特点,但它们的主要区别在于设计理念、功能以及与Kubernetes的集成程度。

1. 设计理念

- ArgoCD:采用声明式方法管理Kubernetes应用程序,通过监控Git仓库的状态变化来触发更新。
- Jenkins:基于事件驱动,通过构建流水线(Pipeline)来实现自动化构建、测试和部署过程。

2. 功能

- ArgoCD:专注于简化从源代码到生产环境的无缝集成,提供应用同步、回滚等功能,并支持Webhooks以减少轮询延迟。

- **Jenkins**：提供丰富的插件生态系统，支持多种编程语言和框架，能够执行复杂的构建任务，包括Maven打包、Docker构建等。

3. 集成Kubernetes

- **ArgoCD**：原生支持Kubernetes，利用Kubernetes的扩展能力（如自动缩放、并发处理等），为云原生应用提供高效的CI/CD解决方案。
- **Jenkins**：虽然可以通过插件与Kubernetes集成，但并非专为Kubernetes设计，可能需要额外的配置和脚本来管理Kubernetes资源。

总的来说，ArgoCD和Jenkins都是强大的CI/CD工具，选择哪个取决于具体的项目需求和技术栈。如果项目深度依赖Kubernetes环境，追求声明式管理和自动化同步功能，ArgoCD可能是更好的选择。而如果需要更广泛的插件支持和更复杂的构建流程管理，Jenkins则可能更适合。

12.2 Kubernetes部署ArgoCD

在Kubernetes集群中部署ArgoCD的步骤如下：

01 在k8s-master节点创建命名空间：

```
[root@k8s-master ~]# kubectl create namespace argocd
namespace/argocd created
```

02 在k8s-master节点安装ArgoCD：

```
[root@k8s-master ~]# kubectl apply -n argocd -f
https://raw.githubusercontent.com/argoproj/argo-cd/stable/manifests/install.yaml
customresourcedefinition.apiextensions.k8s.io/applications.argoproj.io created
customresourcedefinition.apiextensions.k8s.io/applicationsets.argoproj.io created
customresourcedefinition.apiextensions.k8s.io/appprojects.argoproj.io created
serviceaccount/argocd-application-controller created
serviceaccount/argocd-applicationset-controller created
serviceaccount/argocd-dex-server created
serviceaccount/argocd-notifications-controller created
serviceaccount/argocd-redis created
serviceaccount/argocd-repo-server created
serviceaccount/argocd-server created
role.rbac.authorization.k8s.io/argocd-application-controller created
role.rbac.authorization.k8s.io/argocd-applicationset-controller created
role.rbac.authorization.k8s.io/argocd-dex-server created
role.rbac.authorization.k8s.io/argocd-notifications-controller created
role.rbac.authorization.k8s.io/argocd-redis created
role.rbac.authorization.k8s.io/argocd-server created
clusterrole.rbac.authorization.k8s.io/argocd-application-controller created
clusterrole.rbac.authorization.k8s.io/argocd-applicationset-controller created
clusterrole.rbac.authorization.k8s.io/argocd-server created
```

```
rolebinding.rbac.authorization.k8s.io/argocd-application-controller created
rolebinding.rbac.authorization.k8s.io/argocd-applicationset-controller created
rolebinding.rbac.authorization.k8s.io/argocd-dex-server created
rolebinding.rbac.authorization.k8s.io/argocd-notifications-controller created
rolebinding.rbac.authorization.k8s.io/argocd-redis created
rolebinding.rbac.authorization.k8s.io/argocd-server created
clusterrolebinding.rbac.authorization.k8s.io/argocd-application-controller created
clusterrolebinding.rbac.authorization.k8s.io/argocd-applicationset-controller created
clusterrolebinding.rbac.authorization.k8s.io/argocd-server created
configmap/argocd-cm created
configmap/argocd-cmd-params-cm created
configmap/argocd-gpg-keys-cm created
configmap/argocd-notifications-cm created
configmap/argocd-rbac-cm created
configmap/argocd-ssh-known-hosts-cm created
configmap/argocd-tls-certs-cm created
secret/argocd-notifications-secret created
secret/argocd-secret created
service/argocd-applicationset-controller created
service/argocd-dex-server created
service/argocd-metrics created
service/argocd-notifications-controller-metrics created
service/argocd-redis created
service/argocd-repo-server created
service/argocd-server created
service/argocd-server-metrics created
deployment.apps/argocd-applicationset-controller created
deployment.apps/argocd-dex-server created
deployment.apps/argocd-notifications-controller created
deployment.apps/argocd-redis created
deployment.apps/argocd-repo-server created
deployment.apps/argocd-server created
statefulset.apps/argocd-application-controller created
networkpolicy.networking.k8s.io/argocd-application-controller-network-policy created
networkpolicy.networking.k8s.io/argocd-applicationset-controller-network-policy created
networkpolicy.networking.k8s.io/argocd-dex-server-network-policy created
networkpolicy.networking.k8s.io/argocd-notifications-controller-network-policy created
networkpolicy.networking.k8s.io/argocd-redis-network-policy created
networkpolicy.networking.k8s.io/argocd-repo-server-network-policy created
networkpolicy.networking.k8s.io/argocd-server-network-policy created
```

03 该命令将从指定的URL下载并应用配置文件，这里是Argo CD官方提供安装的YAML文件。指定命令操作的命名空间为argocd。

根据输出信息，可以看出已经成功创建了以下资源：

```
customresourcedefinition.apiextensions.k8s.io/applications.argoproj.io created
customresourcedefinition.apiextensions.k8s.io/applicationsets.argoproj.io created
customresourcedefinition.apiextensions.k8s.io/appprojects.argoproj.io created
serviceaccount/argocd-application-controller created
serviceaccount/argocd-applicationset-controller created
serviceaccount/argocd-dex-server created
serviceaccount/argocd-notifications-controller created
serviceaccount/argocd-redis created
serviceaccount/argocd-repo-server created
serviceaccount/argocd-server created
role.rbac.authorization.k8s.io/argocd-application-controller created
role.rbac.authorization.k8s.io/argocd-applicationset-controller created
role.rbac.authorization.k8s.io/argocd-dex-server created
role.rbac.authorization.k8s.io/argocd-notifications-controller created
role.rbac.authorization.k8s.io/argocd-redis created
role.rbac.authorization.k8s.io/argocd-server created
clusterrole.rbac.authorization.k8s.io/argocd-application-controller created
clusterrole.rbac.authorization.k8s.io/argocd-applicationset-controller created
clusterrole.rbac.authorization.k8s.io/argocd-server created
rolebinding.rbac.authorization.k8s.io/argocd-application-controller created
rolebinding.rbac.authorization.k8s.io/argocd-applicationset-controller created
rolebinding.rbac.authorization.k8s.io/argocd-dex-server created
rolebinding.rbac.authorization.k8s.io/argocd-notifications-controller created
rolebinding.rbac.authorization.k8s.io/argocd-redis created
rolebinding.rbac.authorization.k8s.io/argocd-server created
clusterrolebinding.rbac.authorization.k8s.io/argocd-application-controller created
clusterrolebinding.rbac.authorization.k8s.io/argocd-applicationset-controller created
clusterrolebinding.rbac.authorization.k8s.io/argocd-server created
configmap/argocd-cm created
configmap/argocd-cmd-params-cm created
configmap/argocd-gpg-keys-cm created
configmap/argocd-notifications-cm created
configmap/argocd-rbac-cm created
configmap/argocd-ssh-known-hosts-cm created
configmap/argocd-tls-certs-cm created
secret/argocd-notifications-secret created
secret/argocd-secret created
service/argocd-applicationset-controller created
service/argocd-dex-server created
service/argocd-metrics created
service/argocd-notifications-controller-metrics created
service/argocd-redis created
service/argocd-repo-server created
service/argocd-server created
service/argocd-server-metrics created
deployment.apps/argocd-applicationset-controller created
deployment.apps/argocd-dex-server created
deployment.apps/argocd-notifications-controller created
deployment.apps/argocd-redis created
```

```
    deployment.apps/argocd-repo-server created
    deployment.apps/argocd-server created
    statefulset.apps/argocd-application-controller created
    networkpolicy.networking.k8s.io/argocd-application-controller-network-policy
created
    networkpolicy.networking.k8s.io/argocd-applicationset-controller-network-policy
created
    networkpolicy.networking.k8s.io/argocd-dex-server-network-policy created
    networkpolicy.networking.k8s.io/argocd-notifications-controller-network-policy
created
    networkpolicy.networking.k8s.io/argocd-redis-network-policy created
    networkpolicy.networking.k8s.io/argocd-repo-server-network-policy created
    networkpolicy.networking.k8s.io/argocd-server-network-policy created
```

04 查看ArgoCD服务各资源的状态，如图12-1所示。

```
[root@k8s-master ~]# kubectl get all -n argocd
NAME                                                     READY   STATUS    RESTARTS   AGE
pod/argocd-application-controller-0                      1/1     Running   0          9m27s
pod/argocd-applicationset-controller-6c9ffcb978-9z9pf    1/1     Running   0          9m29s
pod/argocd-dex-server-dbccdf8d8-grdwv                    1/1     Running   0          9m29s
pod/argocd-notifications-controller-6d7b75d9b-rln7q      1/1     Running   0          9m29s
pod/argocd-redis-78ff6464dd-6k9ct                        1/1     Running   0          9m29s
pod/argocd-repo-server-6d9474d5c6-4rhbq                  1/1     Running   0          9m28s
pod/argocd-server-5cdbb4db49-9vmxc                       1/1     Running   0          9m27s

NAME                                              TYPE        CLUSTER-IP       EXTERNAL-IP   PORT(S)                      AGE
service/argocd-applicationset-controller          ClusterIP   10.98.84.12      <none>        7000/TCP,8080/TCP            9m34s
service/argocd-dex-server                         ClusterIP   10.105.70.105    <none>        5556/TCP,5557/TCP,5558/TCP   9m34s
service/argocd-metrics                            ClusterIP   10.109.107.100   <none>        8082/TCP                     9m34s
service/argocd-notifications-controller-metrics   ClusterIP   10.98.119.116    <none>        9001/TCP                     9m34s
service/argocd-redis                              ClusterIP   10.109.190.25    <none>        6379/TCP                     9m32s
service/argocd-repo-server                        ClusterIP   10.103.235.187   <none>        8081/TCP,8084/TCP            9m32s
service/argocd-server                             ClusterIP   10.99.124.204    <none>        80/TCP,443/TCP               9m32s
service/argocd-server-metrics                     ClusterIP   10.101.25.130    <none>        8083/TCP                     9m31s

NAME                                               READY   UP-TO-DATE   AVAILABLE   AGE
deployment.apps/argocd-applicationset-controller   1/1     1            1           9m31s
deployment.apps/argocd-dex-server                  1/1     1            1           9m31s
deployment.apps/argocd-notifications-controller    1/1     1            1           9m30s
deployment.apps/argocd-redis                       1/1     1            1           9m29s
deployment.apps/argocd-repo-server                 1/1     1            1           9m29s
deployment.apps/argocd-server                      1/1     1            1           9m28s

NAME                                                          DESIRED   CURRENT   READY   AGE
replicaset.apps/argocd-applicationset-controller-6c9ffcb978   1         1         1       9m31s
replicaset.apps/argocd-dex-server-dbccdf8d8                   1         1         1       9m31s
replicaset.apps/argocd-notifications-controller-6d7b75d9b     1         1         1       9m29s
replicaset.apps/argocd-redis-78ff6464dd                       1         1         1       9m29s
replicaset.apps/argocd-repo-server-6d9474d5c6                 1         1         1       9m28s
replicaset.apps/argocd-server-5cdbb4db49                      1         1         1       9m28s

NAME                                             READY   AGE
statefulset.apps/argocd-application-controller   1/1     9m27s
```

图12-1　ArgoCD 服务各资源的状态

05 查看ArgoCD密码：

```
[root@k8s-master ~]# kubectl get secret --namespace argocd
argocd-initial-admin-secret -o jsonpath="{.data.password}" | base64 --decode ; echo
# 输出信息
9rfED5btLh9qji9E
```

该命令的作用是从Kubernetes集群中检索Argo CD的初始管理员密码，并将其解码后输出。

06 修改Service的类型，将Cluster更改为NodePort：

```
[root@k8s-master ~]# kubectl patch svc argocd-server -p '{"spec":{"type":"NodePort"}}'
 -n argocd
 # 输出信息
 service/argocd-server patched
```

该命令用于更新Kubernetes集群中argocd命名空间下名为argocd-server的Service对象，将其类型从当前的ClusterIP更改为NodePort。这意味着更改后，argocd-server服务的端口将被暴露在每个集群节点的指定端口上，使得用户可以通过任何节点的IP地址和该端口来访问该服务，而不仅仅是集群内部。

07 查看Service的NodePort端口：

```
[root@k8s-master ~]# kubectl get svc -n argocd | grep argocd-server | grep -v
argocd-server-metrics | awk '{print $5}'
 # 输出信息
 80:23012/TCP,443:38080/TCP
```

通过节点IP以及NodePort的端口进行访问，如图12-2所示，此处访问的是443的映射端口，因为argocd会自动跳转到HTTPS，所以需要访问443的映射端口，地址为https://192.168.1.11:38080。

图 12-2　ArgoCD 登录页面

08 账号默认为admin，密码为上面通过命令查询到的9rfED5btLh9qji9E，登录后如图12-3所示。

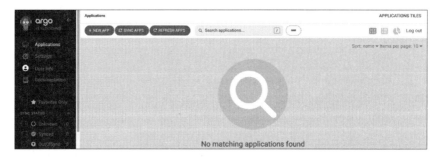

图 12-3　ArgoCD 登录成功首页

12.3 ArgoCD的配置及使用

本节将深入探讨ArgoCD的配置及使用，特别是如何将其与Kubernetes集群连接以及如何使用ArgoCD CLI工具来集成GitLab并创建应用。

12.3.1 ArgoCD 连接 Kubernetes

本小节将学习如何将ArgoCD与现有的Kubernetes集群进行连接，以确保两者可以无缝协作。具体的操作步骤如下：

01 首次登录系统后，依次打开Settings→Clusters，查看默认的连接信息，如图12-4所示。

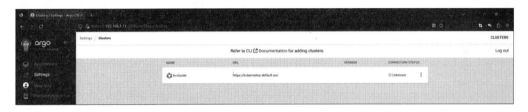

图 12-4　配置集群列表页面

02 修改其中的NAME后，单击INVALIDATE CACHE按钮，查看状态为成功，如图12-5所示。

图 12-5　配置集群成功

12.3.2 使用 ArgoCD CLI 集成 GitLab 并创建 App

本小节将通过实际操作演示如何使用ArgoCD的命令行界面（Command-Line Interface，CLI）工具来集成流行的版本控制系统GitLab，并介绍如何创建和管理应用程序。

01 首先需要安装ArgoCD CLI工具。本示例中，我们将在k8s-master节点上安装此工具：

```
[root@k8s-master ~]# wget
https://github.com/argoproj/argo-cd/releases/download/v2.5.2/argocd-linux-amd64
```

该命令从指定的https://github.com/argoproj/argo-cd/releases/download/v2.5.2/argocd-linux-amd64 URL下载一个名为argocd-linux-amd64的文件，并将其保存到当前目录中。如果下载成功，则该文件将被保存到当前目录下。

02 复制到软件存放目录：

```
[root@k8s-master ~]# cp argocd-linux-amd64 /usr/local/bin/argocd
```

03 赋值可执行权限：

```
[root@k8s-master ~]# chmod +x /usr/local/bin/argocd
```

该命令的作用是为/usr/local/bin/argocd文件添加执行权限。chmod是一个在UNIX、Linux和类UNIX系统中用于改变文件或目录权限的命令。权限决定了谁可以读取、写入或执行文件。+x选项指定了要添加（+）执行（x）权限。执行权限允许用户（或其他用户，取决于权限设置）运行该文件作为程序。

04 查看版本以验证安装是否成功：

```
[root@k8s-master ~]# argocd version
argocd: v2.5.2+148d8da
  BuildDate: 2022-11-07T17:06:04Z
  GitCommit: 148d8da7a996f6c9f4d102fdd8e688c2ff3fd8c7
  GitTreeState: clean
  GoVersion: go1.18.7
  Compiler: gc
  Platform: linux/amd64
argocd-server: v2.11.3+3f344d5
  BuildDate: 2024-06-06T08:42:00Z
  GitCommit: 3f344d54a4e0bbbb4313e1c19cfe1e544b162598
  GitTreeState: clean
  GoVersion: go1.21.9
  Compiler: gc
  Platform: linux/amd64
  Kustomize Version: v5.2.1 2023-10-19T20:13:51Z
  Helm Version: v3.14.4+g81c902a
  Kubectl Version: v0.26.11
  Jsonnet Version: v0.20.0
```

05 通过CLI工具命令行登录ArgoCD：

```
[root@k8s-master ~]# argocd login 192.168.1.11:38080 --username admin --password 9rfED5btLh9qji9E
WARNING: server certificate had error: x509: cannot validate certificate for 192.168.1.11 because it doesn't contain any IP SANs. Proceed insecurely (y/n)? y
'admin:login' logged in successfully
Context '192.168.1.11:38080' updated
```

该命令用于登录Argo CD（一个开源的GitOps持续交付工具）。这个命令尝试使用指定的IP地址（192.168.1.11）、端口（38080）、用户名（admin）和密码（9rfED5btLh9qji9E）来登录Argo CD服务器。

06 在GitLab上创建名为myapp的项目仓库并创建资源清单文件：

```
apiVersion: apps/v1
kind: Deployment
metadata:
  name: myapp
  labels:
    app: myapp
spec:
  replicas: 3
  selector:
    matchLabels:
      app: myapp
  template:
    metadata:
      labels:
        app: myapp
    spec:
      containers:
      - name: my-container
        image: registry.cn-hangzhou.aliyuncs.com/deanmr/dean:myappv1
        ports:
        - containerPort: 80
---
kind: Service
apiVersion: v1
metadata:
  name: myapp
  labels:
    name: myapp
spec:
  ports:
    - name: http
      protocol: TCP
      port: 80
      targetPort: http
  selector:
    name: myapp
  type: ClusterIP
```

07 创建仓库成功后，如图12-6所示，仓库名为myapp，在myapp目录下存在myapp.yaml资源清单。

08 在k8s-master节点创建命名空间：

```
[root@k8s-master ~]# kubectl create ns myapp-uat
```

09 在k8s-master节点创建App资源清单：

图 12-6　仓库下的资源清单信息

```
[root@k8s-master ~]# vim myapp-uat.yaml
apiVersion: argoproj.io/v1alpha1
kind: Application
metadata:
  # App名字
  name: myapp-uat
  # 命名空间必须与ArgoCD的命名空间相匹配，通常为argoCD
  namespace: argocd
spec:
  destination:
    # 部署应用的命名空间
namespace: myapp-uat
# K8s集群内部地址
server: https://kubernetes.default.svc
# 项目名
project: default
  source:
# 资源文件路径，在GitLab仓库中的目录
path: myapp
# Git仓库地址
repoURL: http://gitlab.deanit.cn/root/myapp.git
# 分支名
    targetRevision: master
```

10 应用资源：

```
[root@k8s-master ~]# kubectl apply -f myapp-uat.yaml
application.argoproj.io/myapp-uat created
```

⑪ 登录ArgoCD并查看App状态，如图12-7所示，可以看出状态为丢失和不同步。

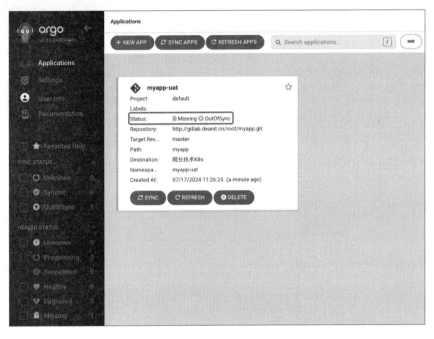

图 12-7　App 状态信息

⑫ 单击图12-7中App的SYNC按钮，进入App详情页后，可以看到GitLab代码仓库中的myapp.yaml文件正在被同步应用于K8s集群中，创建完成后的界面如图12-8所示。

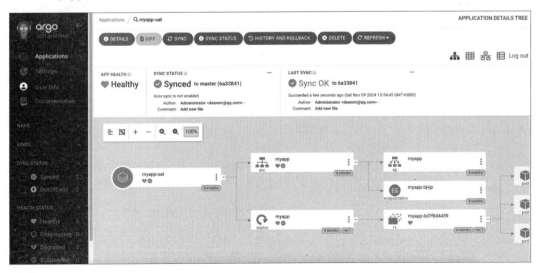

图 12-8　App 同步资源详情页

⑬ 查看myapp的资源状态，可以看出资源清单已经被应用并创建：

```
[root@k8s-master ~]# kubectl get all -n myapp-uat
NAME                               READY   STATUS    RESTARTS   AGE
pod/myapp-6d7f8d4459-862xs         1/1     Running   0          99s
pod/myapp-6d7f8d4459-blkr7         1/1     Running   0          98s
pod/myapp-6d7f8d4459-hv4gw         1/1     Running   0          98s

NAME            TYPE        CLUSTER-IP       EXTERNAL-IP   PORT(S)   AGE
service/myapp   ClusterIP   10.108.208.204   <none>        80/TCP    100s

NAME                    READY   UP-TO-DATE   AVAILABLE   AGE
deployment.apps/myapp   3/3     3            3           100s

NAME                              DESIRED   CURRENT   READY   AGE
replicaset.apps/myapp-6d7f8d4459   3         3         3       99s
```

14 通过ArgoCD CLI工具查看App列表：

```
[root@k8s-master ~]# argocd app list
NAME               CLUSTER                         NAMESPACE  PROJECT  STATUS  HEALTH   SYNCPOLICY  CONDITIONS  REPO                                      PATH   TARGET
argocd/myapp-uat   https://kubernetes.default.svc  myapp-uat  default  Synced  Healthy  <none>      <none>      http://gitlab.deanit.cn/root/myapp.git    myapp  master
```

15 通过ArgoCD CLI工具查看App信息：

```
[root@k8s-master ~]# argocd app get myapp-uat
Name:             argocd/myapp-uat
Project:          default
Server:           https://kubernetes.default.svc
Namespace:        myapp-uat
URL:              https://192.168.1.11:38080/applications/myapp-uat
Repo:             http://gitlab.deanit.cn/root/myapp.git
Target:           master
Path:             myapp
SyncWindow:       Sync Allowed
Sync Policy:      <none>
Sync Status:      Synced to master (6a33841)
Health Status:    Healthy

GROUP  KIND        NAMESPACE  NAME   STATUS  HEALTH   HOOK  MESSAGE
       Service     myapp-uat  myapp  Synced  Healthy        service/myapp created
apps   Deployment  myapp-uat  myapp  Synced  Healthy        deployment.apps/myapp created
```

常用命令：

```
argocd app actions - 管理资源的行为
argocd app create - 创建一个应用程序
argocd app delete - 删除一个应用程序
argocd app delete-resource - 删除应用中的资源
argocd app diff -对目标和活动状态执行差异
```

```
argocd app edit - 编辑应用程序
argocd app get - 获取应用程序的细节
argocd app history - 显示应用程序部署历史
argocd app list - 应用程序列表
argocd app logs - 获取应用程序舱的日志
argocd app manifests - 打印应用程序的清单
argocd app patch - 补丁程序
argocd app patch-resource - 应用程序中的补丁资源
argocd app resources - 列出申请资源
argocd app rollback - 根据历史ID回滚应用程序到以前部署的版本,省略将回滚到以前的版本
argocd app set - 设置应用程序参数
argocd app sync - 将应用程序同步到目标状态
argocd app terminate-op - 终止应用程序的运行操作
argocd app unset - 设置应用程序参数
argocd app wait - 等待应用程序达到同步且正常的状态
```

通过这些步骤,读者将能够搭建起一个功能全面的持续交付系统,从而简化Kubernetes应用程序的部署和管理过程。

12.4 本章小结

本章介绍了如何在Kubernetes集群中安装ArgoCD,实现应用部署的自动化与版本控制。首先,我们详细阐述了如何在Kubernetes集群中安装和配置ArgoCD。这一过程包括准备必要的命名空间,使用官方提供资源清单简化安装步骤。随后,我们深入探讨了如何将ArgoCD与Kubernetes集群进行连接。这一步骤是ArgoCD能够监控并同步集群状态的关键。通过配置ArgoCD的集群管理功能,用户可以添加多个Kubernetes集群作为目标集群,实现跨集群的应用部署和管理。为了进一步提升开发到生产环境的自动化水平,本章还重点介绍了如何使用ArgoCD CLI集成GitLab。通过这一集成,ArgoCD能够自动监听GitLab仓库中的代码变更,并根据这些变更自动更新Kubernetes集群中的应用状态。

综上所述,本章通过翔实的步骤和示例,全面介绍了在Kubernetes集群中安装和配置ArgoCD,以及如何将ArgoCD与GitLab集成以实现自动化部署的过程。这些知识和技能对于希望提高应用部署效率、实现持续交付的开发者和管理员来说具有重要的参考价值。

第 13 章 云原生负载均衡之MetalLB

在当今的云原生环境中，负载均衡是确保应用高可用性和可扩展性的关键组件。MetalLB作为一种开源的负载均衡实现方案，为Kubernetes集群提供了一种简便的方式来暴露服务到外部网络。本章将深入探讨MetalLB的配置与使用，帮助读者掌握如何在Kubernetes环境中部署和管理负载均衡器。

13.1 自建LoadBalancer种类

目前常用的自建LoadBalancer共有三种，如果在裸机环境中运行Kubernetes集群，并且需要对外暴露LoadBalancer服务，那么MetalLB、PureLB和OpenELB都是不错的选择。然而，它们在配置方式、路由协议支持、流量分散策略等方面有所不同，需要根据自己的具体需求来选择最合适的解决方案。

1. MetalLB

MetalLB是一个专为裸机Kubernetes集群设计的负载均衡器，旨在在没有云提供商托管负载均衡服务的环境中工作。它通过ARP、NDP或BGP等协议将集群内部的LoadBalancer类型服务暴露给外部网络。

MetalLB的主要特点如下。

- 虚拟网卡：PureLB会在Kubernetes集群的受管节点上新建一个虚拟网卡（如kube-lb0），用于处理LoadBalancer VIP的流量。
- 路由协议灵活性：PureLB支持使用任意路由协议（如BGP、OSPF等）实现ECMP，这使其在网络配置上更加灵活。
- 流量分散：PureLB的Layer2模式会根据单个VIP来选择节点，从而将多个VIP分散到集群中的不同节点上，以平衡流量。

- 配置方式：PureLB使用CRD（Custom Resource Definition，自定义资源定义）进行配置，但与MetalLB相比，其实现方式和配置细节可能有所不同。

2. PureLB

PureLB是另一种Kubernetes负载均衡器实现，它提供了类似于MetalLB的功能，但在某些方面有所不同，特别是其Layer2模式和ECMP模式的实现。

PureLB的特点如下。

- 虚拟网卡：PureLB会在Kubernetes集群的受管节点上新建一个虚拟网卡（如kube-lb0），用于处理LoadBalancer VIP的流量。
- 路由协议的灵活性：PureLB支持使用任意路由协议（如BGP、OSPF等）实现ECMP，这使其在网络配置上更加灵活。
- 流量分散：PureLB的Layer2模式会根据单个VIP来选择节点，从而将多个VIP分散到集群中的不同节点上，以平衡流量。
- 配置方式：PureLB也使用CRD（自定义资源定义）进行配置，但与MetalLB相比，其实现方式和配置细节可能有所不同。

3. OpenELB

OpenELB是一个由KubeSphere团队开源的负载均衡器插件，旨在为物理机、边缘和私有化环境提供LoadBalancer服务暴露能力。它是PorterLB的后续项目，并已被CNCF沙箱（Sandbox）托管。

OpenELB的特点如下。

- BGP与Layer 2模式：OpenELB支持基于BGP和Layer 2模式的负载均衡，允许用户根据网络环境选择合适的模式。
- IP地址池管理：OpenELB提供了IP地址池管理功能，允许用户通过CRD配置和管理IP地址。
- 易用性：OpenELB的设计考虑到了易用性，用户可以通过Kubectl或Helm Chart等方式轻松部署和管理。
- 社区支持：作为CNCF沙箱项目，OpenELB得到了云原生社区的支持和关注，未来可能会迭代出更多功能和改进。

13.2 MetalLB的核心概念与架构

本节将介绍MetalLB的关键组件、术语和架构，帮助读者理解其工作方式。

13.2.1 MetalLB 的核心概念

1. 负载均衡器

MetalLB在Kubernetes集群中部署，它作为外部流量的入口点，负责将外部流量动态地分发到集群中的服务。

2. 服务

在Kubernetes中，服务是一组提供相同功能的Pod的抽象。通过服务可以实现负载均衡、服务发现和路由等功能。MetalLB主要关注如何将外部流量引导到这些服务上。

3. 负载均衡模式

MetalLB支持两种主要的负载均衡模式：Layer 2模式和BGP模式。

- Layer 2模式：通过ARP协议在本地网络中广播ARP包来绑定IP地址和MAC地址，实现负载均衡。此模式适用于本地网络环境。
- BGP模式：通过BGP协议与网络设备进行通信，将流量路由到Kubernetes集群中的服务。此模式适用于云服务提供商的环境或需要跨网络边界进行路由的场景。

4. IP池管理

MetalLB管理一个IP池，用于分配给需要进行外部访问的服务。管理员可以定义IP池的范围，并确保它们与集群的网络环境兼容。

13.2.2 MetalLB 架构

MetalLB的架构主要由两个组件组成：Controller和Speaker。

1. Controller

- 以Deployment方式部署在Kubernetes集群中。
- 负责监听Kubernetes集群中Service资源的变化。
- 当Service配置为LoadBalancer模式时，Controller会从IP池中分配一个IP地址给该Service，并进行IP地址的生命周期管理。

2. Speaker

- 以DaemonSet方式部署在Kubernetes集群的每个节点上。
- 根据Service的变化，按具体的协议（ARP或BGP）发起相应的广播或应答。
- 在Layer 2模式下，Speaker会广播ARP包来绑定Service的IP地址和集群中某个节点的MAC地址。

- 在BGP模式下，Speaker会与网络设备建立BGP对等会话，并通告外部服务的IP地址，以便网络设备能够将流量路由到正确的节点。

13.3 Kubernetes部署MetalLB

本节将详细介绍如何在Kubernetes环境中部署和使用MetalLB，包括必要的前提条件和步骤。

在部署MetalLB服务之前还需要先了解什么是IPVS（IP Virtual Server）。IPVS是Linux内核自带的一个实现虚拟服务器的软件，同时也是运行在LVS（Linux Virtual Server）下的提供负载平衡功能的一种技术。

13.3.1 检查是否开启 IPVS 功能

本小节首先介绍系统是否已经启用了IPVS功能，这是MetalLB正常工作所需的关键依赖项。

具体操作步骤如下：

01 检查kube-proxy是否启用IPVS：

```
[root@k8s-master ~]# kubectl get configmap kube-proxy -n kube-system -o yaml | grep mode
# 输出信息
    mode: "ipvs"
```

02 检查IPVS内核模块是否加载：

```
[root@k8s-master ~]# lsmod | grep ip_vs
ip_vs_sh               12288  0
ip_vs_wrr              12288  0
ip_vs_rr               12288  64
ip_vs                  200704  70 ip_vs_rr,ip_vs_sh,ip_vs_wrr
nf_conntrack           188416  6 xt_conntrack,nf_nat,xt_nat,nf_conntrack_netlink,xt_MASQUERADE,ip_vs
nf_defrag_ipv6         24576  2 nf_conntrack,ip_vs
libcrc32c              12288  5 nf_conntrack,nf_nat,nf_tables,xfs,ip_vs
```

03 如果在IPVS模式下使用kube-proxy，从Kubernetes v1.14.2开始，必须启用严格的ARP模式。
请注意，如果选择使用kube-router作为服务代理，则无须额外配置，因为它默认启用严格ARP。

```
[root@k8s-master ~]# kubectl edit configmap -n kube-system kube-proxy
apiVersion: kubeproxy.config.k8s.io/v1alpha1
kind: KubeProxyConfiguration
mode: "ipvs"
ipvs:
  strictARP: true  # 设置为true，即开启
```

04 删除kube-system的Pod使之ConfigMap配置生效：

```
[root@k8s-master ~]# kubectl get pods -n kube-system | grep kube-proxy | awk '{print
$1}' | xargs kubectl delete pods -n kube-system
```

13.3.2 配置并创建 MetaLB 服务

接下来，我们将通过实际操作演示如何在Kubernetes中配置和创建MetaLB服务，确保读者能够顺利地设置自己的负载均衡器。

01 在k8s-master节点创建命名空间：

```
[root@k8s-master ~]# kubectl apply -f
https://cdn.jsdelivr.net/gh/metallb/metallb@v0.10/manifests/namespace.yaml
namespace/metallb-system created
```

该命令将从指定的URL下载并应用配置文件到Kubernetes集群中，使用-f选项指定要应用的文件名。根据输出信息，可以看出已经成功创建了以下资源：

```
namespace/metallb-system created
```

02 在k8s-master节点创建并应用Layer2配置资源清单：

```
[root@k8s-master ~]# kubectl apply -f - <<EOF
apiVersion: v1
kind: ConfigMap
metadata:
  namespace: metallb-system
  name: config
data:
  config: |
    address-pools:
    - name: default
      protocol: layer2
      addresses:
      # 设置LB自动分配的IP范围，切记不要把K8s集群节点IP包含进去，避免IP冲突
      - 192.168.1.20-192.168.1.50
EOF
```

03 应用Metallb资源清单，并替换资源清单中的镜像地址：

```
[root@k8s-master ~]# curl https://cdn.jsdelivr.net/gh/metallb/metallb@v0.10
/manifests/metallb.yaml | sed -e "s/quay.io\///" | kubectl apply -f -
  % Total    % Received % Xferd  Average Speed   Time    Time     Time  Current
                                 Dload  Upload   Total   Spent    Left  Speed
100  8391  100  8391    0     0   5864      0  0:00:01  0:00:01 --:--:--  5863
Warning: policy/v1beta1 PodSecurityPolicy is deprecated in v1.21+, unavailable in
v1.25+
podsecuritypolicy.policy/controller created
podsecuritypolicy.policy/speaker created
serviceaccount/controller created
serviceaccount/speaker created
```

```
clusterrole.rbac.authorization.k8s.io/metallb-system:controller created
clusterrole.rbac.authorization.k8s.io/metallb-system:speaker created
role.rbac.authorization.k8s.io/config-watcher created
role.rbac.authorization.k8s.io/pod-lister created
role.rbac.authorization.k8s.io/controller created
clusterrolebinding.rbac.authorization.k8s.io/metallb-system:controller created
clusterrolebinding.rbac.authorization.k8s.io/metallb-system:speaker created
rolebinding.rbac.authorization.k8s.io/config-watcher created
rolebinding.rbac.authorization.k8s.io/pod-lister created
rolebinding.rbac.authorization.k8s.io/controller created
daemonset.apps/speaker created
deployment.apps/controller created
```

该命令的作用是从metallb的特定版本（v0.10）下载MetallB的Kubernetes配置文件metallb.yaml，然后通过sed命令修改该配置文件中的镜像地址（移除quay.io/前缀），最后使用kubectl apply命令将修改后的配置应用到Kubernetes集群中。根据输出信息，可以看出已经成功创建了以下资源：

```
podsecuritypolicy.policy/controller created
podsecuritypolicy.policy/speaker created
serviceaccount/controller created
serviceaccount/speaker created
clusterrole.rbac.authorization.k8s.io/metallb-system:controller created
clusterrole.rbac.authorization.k8s.io/metallb-system:speaker created
role.rbac.authorization.k8s.io/config-watcher created
role.rbac.authorization.k8s.io/pod-lister created
role.rbac.authorization.k8s.io/controller created
clusterrolebinding.rbac.authorization.k8s.io/metallb-system:controller created
clusterrolebinding.rbac.authorization.k8s.io/metallb-system:speaker created
rolebinding.rbac.authorization.k8s.io/config-watcher created
rolebinding.rbac.authorization.k8s.io/pod-lister created
rolebinding.rbac.authorization.k8s.io/controller created
daemonset.apps/speaker created
deployment.apps/controller created
```

04 查看Metallb的Pod状态：

```
[root@k8s-master ~]# kubectl get pods -n metallb-system
NAME                          READY   STATUS    RESTARTS   AGE
controller-6cb5c4f84c-s5b9x   1/1     Running   0          4m56s
speaker-45w8m                 1/1     Running   0          4m57s
speaker-jm2rk                 1/1     Running   0          4m57s
speaker-np5wh                 1/1     Running   0          4m57s
speaker-rqqbl                 1/1     Running   0          4m57s
```

至此，Metallb服务安装完成。

13.3.3 创建 LoadBalancer 类型的服务

本小节将展示如何使用MetalLB来创建LoadBalancer类型的服务，使得集群内的应用程序可以对外部流量进行响应。

具体操作步骤如下：

01 在k8s-master节点创建资源清单，用于测试Metallb服务是否可以分配IP地址：

```yaml
[root@k8s-master ~]# vim lb-nginx.yaml
apiVersion: apps/v1
kind: Deployment
metadata:
  name: lb-nginx
spec:
  selector:
    matchLabels:
      app: lb-nginx
  template:
    metadata:
      labels:
        app: lb-nginx
    spec:
      containers:
      - name: lb-nginx
        image: nginx:latest
        ports:
        - name: http
          containerPort: 80
---
apiVersion: v1
kind: Service
metadata:
  name: lb-nginx
spec:
  ports:
  - name: http
    port: 80
    protocol: TCP
    targetPort: 80
  selector:
    app: lb-nginx
  type: LoadBalancer   # Service类型为LoadBalancer，将自动从分配一个外部地址
```

02 应用资源：

```
[root@k8s-master ~]# kubectl apply -f lb-nginx.yaml
deployment.apps/lb-nginx created
service/lb-nginx created
```

该命令使用Kubectl工具将lb-nginx.yaml文件中定义的资源应用到Kubernetes集群中。使用-f选项指定要应用的文件名。根据输出信息，可以看出已经成功创建了以下资源：

```
replicaset.apps/rs-nginx-pod created
```

13.3.4 使用 MetalLB 进行服务的外部访问

本小节将介绍如何配置服务以允许外部流量通过MetalLB进行访问，确保读者能够将内部服务暴露给互联网，实现服务的外部可访问性。

具体操作步骤如下：

01 查看资源状态并访问测试：

```
--- 查看Pod状态
[root@k8s-master ~]# kubectl get pods
NAME                         READY   STATUS    RESTARTS   AGE
lb-nginx-5d9cdb7848-skkb4    1/1     Running   0          51s

--- 查看Service状态
[root@k8s-master ~]# kubectl get svc
NAME         TYPE           CLUSTER-IP       EXTERNAL-IP     PORT(S)        AGE
kubernetes   ClusterIP      10.96.0.1        <none>          443/TCP        74d
lb-nginx     LoadBalancer   10.108.218.198   192.168.1.20    80:35139/TCP   56s

--- 访问Metallb分配的IP地址
[root@k8s-master ~]# curl 192.168.1.20
<!DOCTYPE html>
<html>
<head>
<title>Welcome to nginx!</title>
<style>
html { color-scheme: light dark; }
body { width: 35em; margin: 0 auto;
font-family: Tahoma, Verdana, Arial, sans-serif; }
</style>
</head>
<body>
<h1>Welcome to nginx!</h1>
<p>If you see this page, the nginx web server is successfully installed and
working. Further configuration is required.</p>

<p>For online documentation and support please refer to
<a href="http://nginx.org/">nginx.org</a>.<br/>
Commercial support is available at
<a href="http://nginx.com/">nginx.com</a>.</p>

<p><em>Thank you for using nginx.</em></p>
</body>
```

```
</html>
```

02 通过上述命令输出结果，可以看到通过访问LB地址，可以正确请求到后端服务。

13.4 本章小结

本章介绍了MetalLB是一个开源的负载均衡器实现，它解决了Kubernetes中内置的Service类型LoadBalancer在无法直接访问云提供商负载均衡服务的环境中的限制，通过为集群内的服务分配可路由的IP地址，实现了服务的外部访问。接着，我们深入讲解了如何在Kubernetes集群中安装MetalLB、配置MetalLB的ConfigMap以定义IP地址池以及部署MetalLB的控制器和speaker组件。最后，还介绍了如何测试MetalLB的功能。测试步骤通常包括创建一个使用LoadBalancer类型Service的Kubernetes应用，并验证MetalLB是否正确地为该Service分配了IP地址，以及外部用户是否能够通过该IP地址访问应用。通过执行这些测试，可以确保MetalLB已经成功集成到Kubernetes集群中，并能够为集群内的服务提供外部访问能力。

第 14 章 Helm与Loki-Stack搭建日志监控系统

本章将介绍Helm工具的使用以及它与Prometheus、Loki与Grafana的集成应用。Loki是一个由Grafana Labs开发的开源日志聚合系统,专为云原生环境设计。Loki旨在解决在Kubernetes或其他云原生环境中管理和查询大量日志的难题。它特别优化了性能、成本效益和可扩展性,使其非常适合处理来自微服务架构的复杂日志数据流。本章将介绍如何进行Helm的部署,并带领读者实现Loki-Stack日志监控系统的构建。

14.1 Helm包管理与部署

本节介绍Helm包管理工具的核心概念、组件架构以及Helm3与Helm2之间的差异,同时,还将详细介绍如何在CentOS 7系统上安装Helm3,为后续的仓库管理和Release管理打下坚实的基础。

14.1.1 Helm 概述

Helm是Kubernetes的包管理工具,类似于Linux系统下的apt-get或yum,但专为Kubernetes设计。它能够管理Kubernetes应用程序的生命周期,包括安装、升级、回滚和卸载应用。Helm将Kubernetes应用程序的所有资源和配置信息封装在一个Chart包中,方便分发和部署。

Helm的重要组件如下。

- Chart: Helm的包格式,包含一组定义Kubernetes资源的YAML文件。它描述了如何部署一个应用,包括所需的Deployments、Services、ConfigMaps等资源。

一个Chart通常包含以下文件和目录。

- Chart.yaml: Chart的描述文件,包括Chart的名称、版本、描述等信息。

- values.yaml：包含模板文件中可能引用到的变量值。
- templates/：存放所有Kubernetes模板文件的目录，这些模板文件定义了要部署的资源。
- charts/：可选目录，用于存放当前Chart依赖的其他Chart。
- Helm客户端：用户通过Helm客户端与Kubernetes集群交互，执行安装、升级等操作。
- Helm仓库（Repository）：用于存储和分享Chart包的仓库，用户可以从这些仓库中搜索、下载和安装Chart。

使用Helm进行仓库管理时，使用的命令如下：

- 使用helm repo add命令添加Helm仓库。
- 使用helm repo update命令更新仓库列表，以确保可以访问最新的Chart版本。
- 使用helm search repo命令在仓库中搜索Chart。
- 使用helm install命令部署一个Chart到Kubernetes集群时，Helm会创建一个Release。Release是Chart在集群中的一个实例，它代表了特定版本的应用部署。

可以通过以下命令对Release进行管理：

- 使用helm list命令列出所有已安装的Release。
- 使用helm status命令查看特定Release的状态信息。
- 使用helm upgrade命令更新Release到新的Chart版本。
- 使用helm rollback命令将Release回滚到之前的版本。
- 使用helm uninstall命令删除指定的Release。

14.1.2　CentOS 7 系统安装 Helm3

目前的Helm有Helm3和Helm2两个版本，两者的区别主要在架构和命令上，相比Helm2，Helm3移除了Tiller组件，现在的Helm3客户端可以直接与Kubernetes API服务器通信，减少了复杂性，并提高了安全性。另外，部分命令在Helm3中有所变更，例如删除release的命令由helm delete release-name --purge变更为helm uninstall release-name等。

这里以Helm3为例进行讲解。接下来介绍在CentOS7系统中安装Helm3的方法和步骤。

01 下载Helm压缩包：

```
[root@k8s-master ~]# wget https://get.helm.sh/helm-v3.15.2-linux-amd64.tar.gz
```

02 解压Helm压缩包：

```
[root@k8s-master ~]# tar -xf helm-v3.15.2-linux-amd64.tar.gz
```

03 将Helm执行程序复制到系统可执行目录：

```
[root@k8s-master ~]# cp linux-amd64/helm /usr/local/bin/
```

04 验证Helm版本以确认安装是否成功：

```
[root@k8s-master ~]# helm version
version.BuildInfo{Version:"v3.15.2",
GitCommit:"1a500d5625419a524fdae4b33de351cc4f58ec35", GitTreeState:"clean",
GoVersion:"go1.22.4"}
```

14.2 Loki-Stack部署与实践

Loki-Stack是一个轻量级、高可用的日志聚合系统，专为现代云原生环境设计。它包括Loki、Promtail等组件，通过与Prometheus和Grafana的集成，实现了高效的日志收集、存储和可视化。本节将详细介绍如何通过Helm3部署Loki-Stack，实现外部访问Grafana、日志监控查询、导入仪表盘面板以及配置监控告警。

14.2.1 Loki 与 Loki-Stack 概述

1. Loki介绍

Loki是由Grafana Labs开发的一款轻量级、可扩展的日志聚合和检索系统，专为现代云原生环境设计。它可以与Prometheus和Kubernetes等云原生技术栈紧密集成，使用Prometheus的形式抓取数据，并支持Kubernetes中的DaemonSet方式部署Promtail以收集日志。Loki具有如下特点。

- 水平可扩展性：Loki的设计使其可以轻松地在多个节点上进行水平扩展，以应对大规模的日志数据。
- 多租户支持：Loki支持多租户架构，可以在单个实例中处理来自不同租户的日志数据。
- 高效存储：Loki采用紧凑的索引和压缩算法，相比Elasticsearch可以显著减少存储空间。
- 无全文检索：Loki不对日志内容进行全文索引，而是为每个日志流编制一组标签，从而可以简化操作并降低成本。

2. Loki-Stack的组件

Loki-Stack系统包括如下重要组件。

- Promtail：Promtail是Loki的默认客户端，负责采集并上报日志。它基于Prometheus的服务发现机制，能够自动发现采集目标并给日志流添加标签。
- Loki：作为核心组件，Loki负责接收、存储和查询日志数据。它使用与Prometheus相同的服务发现机制，并将标签添加到日志流中。
- Grafana：Grafana用于展示来自Loki的时间序列数据，支持查询、可视化和报警操作。在Loki技术栈中，Grafana专门用来展示来自Prometheus和Loki的数据源。

运维笔记：在实际工作中，可以根据需要对Loki进行垂直扩容和垂直缩容。当日志存储磁盘告警时，可以选择PVC存储扩容。安装完成后，可以通过Grafana界面访问Loki，进行日志查询和可视化。

> **注意** Prometheus是一个由SoundCloud创建的开源系统监控和警报工具套件，Prometheus通过其组件与架构（如Prometheus Server、Exporters/Instrumented Applications、Pushgateway、Alertmanager和Service Discovery）实现了强大的监控能力。它特别擅长实时记录指标，广泛应用于各种监控场景，包括微服务架构、云基础设施、数据库和业务监控等。关于Prometheus的使用请参考相关资料。

> **注意** Grafana是一个开源的数据可视化与监控平台，允许用户通过图表、图形和面板展示、分析和监控来自不同数据源的数据。其设计旨在帮助用户快速理解复杂的系统和应用程序的性能和健康状况。在IT运维监控、网络监控、能源管理和金融市场分析等多种场景中得到了广泛应用。关于Grafana的使用可以参考相关资料。

14.2.2 Helm3 部署 Loki-Stack

使用Helm3可以简化Loki-Stack的部署过程。通过Helm Chart，我们可以快速安装和配置Loki、Promtail以及其他相关组件，确保它们在Kubernetes集群中正确运行。Helm3部署Loki-Stack的步骤如下：

01 添加Grafana的Helm仓库，设置别名为dean：

```
[root@k8s-master ~]# helm repo add dean https://grafana.github.io/helm-charts
"dean" has been added to your repositories
```

该命令用于向Helm添加一个名为dean的仓库，但其实际指向的是Grafana官方维护的Helm图表（Charts）仓库的URL（https://grafana.github.io/helm-charts）。

02 更新仓库列表：

```
[root@k8s-master ~]# helm repo update
Hang tight while we grab the latest from your chart repositories...
...Successfully got an update from the "dean" chart repository
Update Complete. *Happy Helming!*
```

该命令是Helm命令行工具中的一个命令，用于更新所有已配置的Helm仓库的索引信息。当添加一个或多个Helm仓库之后，这些仓库中的Helm图表列表会存储在本地的一个索引文件中。执行helm repo update命令时，Helm会遍历所有已配置的仓库，并尝试从它们的远程位置下载最新的索引文件。这样，就可以使用helm search repo命令来查找最新的图表了。

03 查看所有已安装的仓库列表：

```
[root@k8s-master ~]# helm repo list
NAME    URL
dean    https://grafana.github.io/helm-charts
```

04 使用Helm工具安装Loki-Stack，并通过命令行指定value文件中的参数配置：

```
[root@k8s-master ~]# helm upgrade --install --namespace=monitoring loki-stack dean/loki-stack --set \
grafana.enabled=true,\
prometheus.enabled=true,\
prometheus.alertmanager.enabled=false,\
grafana.image.repository=docker.m.daocloud.io/grafana/grafana,\
grafana.image.tag=10.3.3,\
grafana.sidecar.image.repository=quay.m.daocloud.io/kiwigrid/k8s-sidecar,\
grafana.sidecar.image.tag=1.19.2,\
grafana.persistence.enabled=true,\
grafana.persistentVolume.enabled=true,\
grafana.persistence.storageClassName=dean-nfs,\
grafana.persistence.size=5Gi,\
prometheus.server.persistence.enabled=true,\
prometheus.server.persistentVolume.enabled=true,\
prometheus.server.persistentVolume.storageClass=dean-nfs,\
prometheus.server.persistentVolume.size=5Gi,\
loki.image.repository=docker.m.daocloud.io/grafana/loki,\
loki.image.tag=2.6.1,\
loki.persistence.enabled=true,\
loki.persistentVolume.enabled=true,\
loki.persistence.storageClassName=dean-nfs,\
loki.persistence.size=5Gi,\
promtail.image.registry=docker.m.daocloud.io,\
promtail.image.repository=grafana/promtail,\
promtail.image.tag=2.9.3,\
prometheus.server.image.repository=quay.m.daocloud.io/prometheus/prometheus,\
prometheus.server.image.tag=v2.41.0,\
prometheus.configmapReload.prometheus.image.repository=docker.m.daocloud.io/jimmidyson/configmap-reload,\
prometheus.configmapReload.prometheus.image.tag=v0.8.0,\
prometheus.kube-state-metrics.image.repository=k8s.m.daocloud.io/kube-state-metrics/kube-state-metrics,\
prometheus.kube-state-metrics.image.tag=v2.8.0
```

该命令用于是在Kubernetes集群中安装或升级名为loki-stack的应用，该应用基于dean仓库中的loki-stack Helm图表。通过在--set选项后面跟键－值对，用于覆盖Loki堆栈Helm图表中关于持久化存储以及镜像的配置。请注意，具体的配置项和它们的默认值将取决于正在使用的Helm图表版本和图表本身的设计。

05 查看monitoring命名空间下所有已安装的Helm图表，如图14-1所示。

```
[root@k8s-master ~]# helm list -n monitoring
NAME        NAMESPACE   REVISION    UPDATED                                 STATUS      CHART               APP VERSION
loki-stack  monitoring  1           2024-07-20 23:31:29.004697546 +0800 CST deployed    loki-stack-2.10.2   v2.9.3
```

图 14-1　monitoring 命名空间已安装的 Helm 图表

06 查看monitoring命名空间下的Pod和控制器资源信息，如图14-2所示。

```
[root@k8s-master ~]# kubectl get all -n monitoring
NAME                                                    READY   STATUS    RESTARTS   AGE
pod/loki-stack-0                                        1/1     Running   0          8m54s
pod/loki-stack-grafana-5697599f99-cnwlt                 2/2     Running   0          8m54s
pod/loki-stack-kube-state-metrics-c46d79cb8-jfgfs       1/1     Running   0          8m54s
pod/loki-stack-prometheus-node-exporter-44bbh           1/1     Running   0          8m54s
pod/loki-stack-prometheus-node-exporter-57956           1/1     Running   0          8m54s
pod/loki-stack-prometheus-node-exporter-6ppgl           1/1     Running   0          8m54s
pod/loki-stack-prometheus-node-exporter-ljrg5           1/1     Running   0          8m54s
pod/loki-stack-prometheus-pushgateway-786c597fcc-9fg69  1/1     Running   0          8m54s
pod/loki-stack-prometheus-server-5589bc9cd4-rmmst       2/2     Running   0          8m54s
pod/loki-stack-promtail-clwk6                           1/1     Running   0          8m54s
pod/loki-stack-promtail-kdd56                           1/1     Running   0          8m54s
pod/loki-stack-promtail-pk5f4                           1/1     Running   0          8m54s
pod/loki-stack-promtail-vs2d8                           1/1     Running   0          8m54s

NAME                                          TYPE        CLUSTER-IP       EXTERNAL-IP   PORT(S)    AGE
service/loki-stack                            ClusterIP   10.98.52.154     <none>        3100/TCP   8m55s
service/loki-stack-grafana                    ClusterIP   10.100.108.140   <none>        80/TCP     8m55s
service/loki-stack-headless                   ClusterIP   None             <none>        3100/TCP   8m55s
service/loki-stack-kube-state-metrics         ClusterIP   10.105.139.232   <none>        8080/TCP   8m55s
service/loki-stack-memberlist                 ClusterIP   None             <none>        7946/TCP   8m55s
service/loki-stack-prometheus-node-exporter   ClusterIP   10.100.248.232   <none>        9100/TCP   8m55s
service/loki-stack-prometheus-pushgateway     ClusterIP   10.106.110.48    <none>        9091/TCP   8m55s
service/loki-stack-prometheus-server          ClusterIP   10.97.58.131     <none>        80/TCP     8m55s

NAME                                                 DESIRED   CURRENT   READY   UP-TO-DATE   AVAILABLE   NODE SELECTOR   AGE
daemonset.apps/loki-stack-prometheus-node-exporter   4         4         4       4            4           <none>          8m54s
daemonset.apps/loki-stack-promtail                   4         4         4       4            4           <none>          8m54s

NAME                                                READY   UP-TO-DATE   AVAILABLE   AGE
deployment.apps/loki-stack-grafana                  1/1     1            1           8m54s
deployment.apps/loki-stack-kube-state-metrics       1/1     1            1           8m54s
deployment.apps/loki-stack-prometheus-pushgateway   1/1     1            1           8m54s
deployment.apps/loki-stack-prometheus-server        1/1     1            1           8m54s

NAME                                                           DESIRED   CURRENT   READY   AGE
replicaset.apps/loki-stack-grafana-5697599f99                  1         1         1       8m54s
replicaset.apps/loki-stack-kube-state-metrics-c46d79cb8        1         1         1       8m54s
replicaset.apps/loki-stack-prometheus-pushgateway-786c597fcc   1         1         1       8m54s
replicaset.apps/loki-stack-prometheus-server-5589bc9cd4        1         1         1       8m54s

NAME                          READY   AGE
statefulset.apps/loki-stack   1/1     8m54s
```

图 14-2　monitoring 命名空间下的 Pod 和控制器资源信息

07 查看monitoring命名空间下PV和PVC的资源信息，如图14-3所示。

```
[root@k8s-master ~]# kubectl get pvc -n monitoring
NAME                           STATUS   VOLUME                                     CAPACITY   ACCESS MODES   STORAGECLASS   AGE
loki-stack-grafana             Bound    pvc-8e21d4c4-8946-4218-b000-3bd27a4d9f93   5Gi        RWO            dean-nfs       13m
loki-stack-prometheus-server   Bound    pvc-ac89f41e-8c36-471d-8799-dcf1d17d4608   5Gi        RWO            dean-nfs       13m
storage-loki-stack-0           Bound    pvc-6f44dbdf-f194-49a4-b228-e98ddce08c11   5Gi        RWO            dean-nfs       13m
[root@k8s-master ~]# kubectl get pv | grep monitoring
pvc-6f44dbdf-f194-49a4-b228-e98ddce08c11   5Gi   RWO   Retain   Bound   monitoring/storage-loki-stack-0           dean-nfs   13m
pvc-8e21d4c4-8946-4218-b000-3bd27a4d9f93   5Gi   RWO   Retain   Bound   monitoring/loki-stack-grafana             dean-nfs   13m
pvc-ac89f41e-8c36-471d-8799-dcf1d17d4608   5Gi   RWO   Retain   Bound   monitoring/loki-stack-prometheus-server   dean-nfs   13m
```

图 14-3　monitoring 命名空间下 PV 和 PVC 的资源信息

14.2.3 外部访问 Grafana

为了方便地从外部网络访问Grafana，我们需要配置Ingress或LoadBalancer。这将允许用户通过互联网安全地连接到Grafana实例，查看和管理日志数据。

具体配置步骤如下：

01 在k8s-master节点创建Grafana的Ingress资源清单：

```
[root@k8s-master ~]# vim grafana-ingress.yaml
apiVersion: networking.k8s.io/v1
kind: Ingress
metadata:
  name: grafana
  namespace: monitoring
spec:
  ingressClassName: nginx
  rules:
  - host: grafana.deanit.cn
    http:
      paths:
      - backend:
          service:
            name: loki-stack-grafana
            port:
              number: 80
        path: /
        pathType: Prefix
```

02 应用资源：

```
[root@k8s-master ~]# kubectl apply -f grafana-ingress.yaml
ingress.networking.k8s.io/grafana created
```

03 查看monitoring命名空间下Ingress的资源信息：

```
[root@k8s-master ~]# kubectl get ingress -n monitoring
NAME      CLASS   HOSTS               ADDRESS           PORTS   AGE
grafana   nginx   grafana.deanit.cn   10.110.206.243    80      13s
```

04 查看Grafana密码，账号默认为admin：

```
[root@k8s-master ~]# kubectl get secret --namespace monitoring loki-stack-grafana -o jsonpath="{.data.admin-password}" | base64 --decode ; echo
# 输出信息
F194Rr7q4zN8TtBRpnTOMh7acQT1aLk01W05wRxC
```

由于我们创建了Ingress资源，因此可以通过域名来访问Grafana服务，各客户端配置如下：

```
# Linux和macOS客户端配置
$ vim /etc/hosts
```

```
192.168.1.11 grafana.deanit.cn

# Windows客户端配置
# hosts文件位于路径：C:\Windows\System32\drivers\etc\hosts。
192.168.1.11 grafana.deanit.cn
```

> **注意** 由于hosts文件属于系统文件，因此在修改之前需要获取管理员权限。

05 通过域名访问后，Grafana登录页面如图14-4所示。

图 14-4　Grafana 登录页面

06 查看数据源信息，在左侧菜单中选择Connections→Data sources，可以看到默认已经配置好的数据源，如图14-5所示。

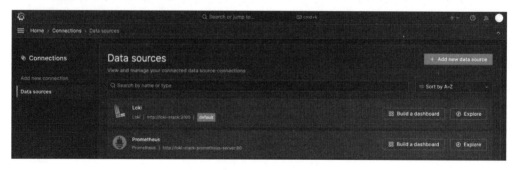

图 14-5　数据源信息

14.2.4　日志监控查询

一旦Loki和Grafana部署完成，我们就可以开始进行日志监控查询。利用Loki的强大查询语法，我们可以轻松过滤和分析日志数据，以发现潜在的问题和异常。

示例：查询app等于ingress-nginx的日志。

在左侧菜单中选择Explore，在Label filters中输入key为app、value为ingress-nginx，单击Run query按钮，进行日志查询，如图14-6所示。

图14-6　日志查询操作

单击 展开日志的详细信息，如图14-7所示。

图14-7　日志的详细信息

14.2.5　导入仪表盘面板

为了更直观地展示日志数据，我们可以导入预定义的仪表盘面板。这些面板提供了丰富的图表和指标，可以帮助我们更好地理解系统的运行状态和性能表现。

具体操作步骤如下：

① 在左侧菜单中选择Dashboards，单击Create Dashboard按钮，如图14-8所示。

图 14-8　创建面板

② 单击Import a dashboard按钮，如图14-9所示。

图 14-9　导入面板

③ 单击Upload dashboard JSON file，选择文件名为KubernetesClusterMonitoring.json的仪表盘面板文件，然后单击"打开"按钮，如图14-10所示。

图 14-10　选择仪表盘文件进行导入

图 14-10　选择仪表盘文件进行导入（续）

04 选择Prometheus数据源，单击Import按钮进行导入，如图14-11所示。

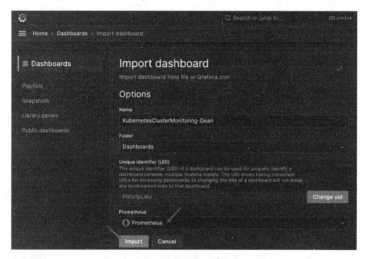

图 14-11　选择数据源进行导入

05 导入成功后，即可看到仪表盘面板，如图14-12所示。

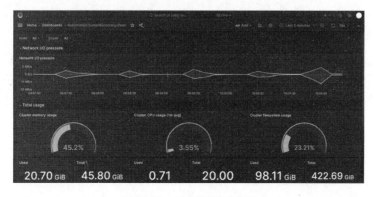

图 14-12　仪表盘展示页面

14.2.6 监控告警

配置监控告警是确保系统稳定运行的重要步骤。通过设置阈值和通知规则,我们可以在检测到异常情况时及时收到警报,从而迅速采取措施解决问题。

具体操作如下:

01 在k8s-master节点修改Grafana的配置文件,添加邮箱支持:

```
[root@k8s-master ~]# kubectl edit cm -n monitoring loki-stack-grafana
# 将以下配置文件添加至ConfigMap
    [smtp]
    enabled = true
    host = smtp.qq.com:465              # 发送服务器地址
    user = deanmr@qq.com                # 你的邮箱
    password = uplsvgphxjtugfjj         # 这个密码是开启smtp服务生成的密码
    from_address = deanmr@qq.com        # 发送邮件的账号
    from_name = dean                    # 自定义的名字
    skip_verify = true                  # 是否跳过验证
    ;ehlo_identity = http://grafana.deanit.cn/     # 域名
    ;cert_file =
    ;key_file =
```

02 删除Grafana的Pod,使之配置生效:

```
[root@k8s-master ~]# kubectl get pods -n monitoring | grep loki-stack-grafana-5697599f99-cnwlt | awk '{print $1}' | xargs kubectl delete pods -n monitoring
```

03 在左侧菜单中选择Alerting→Contact points,然后选择Email,单击Edit按钮,如图14-13所示。

图14-13 Email 配置

04 设置告警介质与收件箱地址,如图14-14所示。

05 在左侧菜单中选择Dashboards,依次单击New→New dashboard新建面板,如图14-15所示。

第 14 章　Helm 与 Loki-Stack 搭建日志监控系统　243

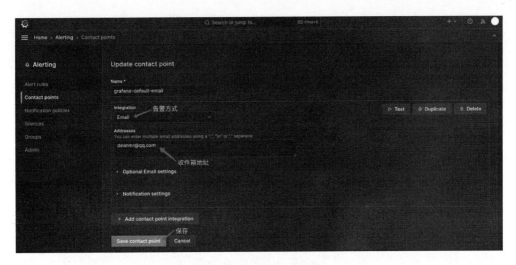

图 14-14　配置 Email 信息

图 14-15　新建面板

06 输入 PromQL，设置仪表盘类型和面板标题，如图 14-16 所示。

```
sum(node_memory_MemAvailable_bytes{instance=~".+"}) by (node)/1024/1024/1024
```

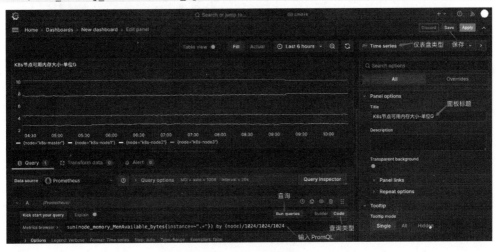

图 14-16　配置面板

07 单击Alert，再单击New alert rule按钮，新建告警规则，如图14-17所示。

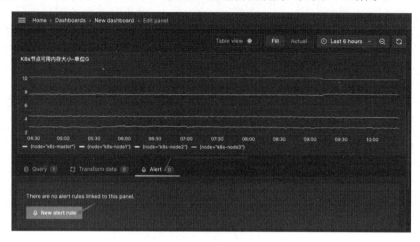

图 14-17　创建告警规则

08 设置告警规则名字，如图14-18所示。

09 定义查询和警报条件，如图14-19所示。

图 14-18　设置告警规则名字

图 14-19　配置告警规则

10 设置评估行为，保存并退出，如图14-20所示。

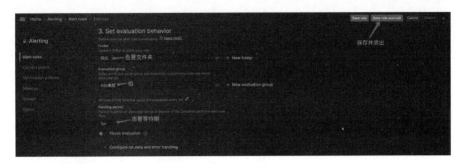

图 14-20　配置评估行为

11 配置告警后，如图14-21所示。

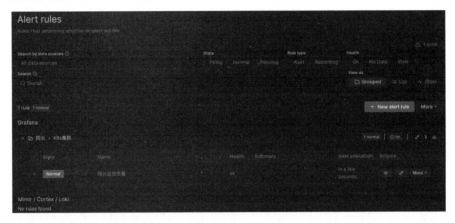

图 14-21　配置告警完成

12 测试邮件告警与恢复，如图14-22所示。

（a）告警

图 14-22　邮件告警触发效果图

(b)恢复

图 14-22　邮件告警触发效果图（续）

14.3　本章小结

本章介绍了如何在Kubernetes环境中利用Helm这个Kubernetes的包管理工具来安装和配置Prometheus、Grafana以及Loki日志系统。首先介绍了Helm包管理与部署，然后简要介绍了Loki，最后通过Helm3部署Loki-Stack实例，介绍了如何在Grafana中配置QQ邮箱告警。本章详细讲解了QQ邮箱告警的配置步骤，包括如何在Grafana中设置SMTP服务器、验证邮箱账户以及创建并测试告警规则，从而实现了对系统异常情况的及时响应和处理，有助于运维人员提升系统的监控和日志管理能力，掌握有效的故障排查和告警通知手段。

第 15 章 Istio微服务时代的服务网格领航者

本章将深入探讨Istio——这一云原生时代的关键技术,Istio被誉为服务网格的领航者。Istio为微服务架构提供了强大的通信管理、流量控制、安全加固及可观测性能力,极大地简化了复杂分布式系统的开发与运维工作。通过学习Istio,读者可以掌握Istio的安装与配置、服务发现与路由管理、安全策略的实施、监控与日志记录的集成等关键使用方法,并通过实践案例展示Istio如何助力企业构建高效、可靠、安全的云原生应用。

15.1 Istio概述

Istio是一个功能强大的服务网格平台,它通过提供负载均衡、服务发现、流量管理、安全性和可观测性等功能,帮助用户更好地管理和控制分布式微服务架构中的网络通信。随着云原生和微服务架构的普及,Istio正逐渐成为构建可靠、高效和安全的微服务应用的重要工具。

Istio服务网格在逻辑上分为数据平面和控制平面。

- 数据平面:由一组智能代理(Envoy)组成,代理部署为边车,用于调解和控制微服务之间所有的网络通信,与控制面进行交互,根据动态更新的配置对代理的微服务的进出流量进行拦截处理,另外生成遥测数据,为微服务提供可观测能力。
- 控制平面:控制平面主要是指Istiod组件,负责管理和配置数据平面中的sidecar代理,Istiod由Pilot、Citadel和Galley三个子组件组成。其中Pilot组件负责提供服务发现、智能路由(如金丝雀发布)和弹性功能(如超时、重试);Citadel负责安全管理,包括密钥和证书的管理;Galley负责配置的验证和处理等功能。

Istio中的流量分为数据平面流量和控制平面流量。

- 数据平面流量：工作负载的业务逻辑发送和接收的消息。
- 控制平面流量：在Istio组件之间发送的配置和控制消息用来编排网格的行为。

Istio架构图如图15-1所示。

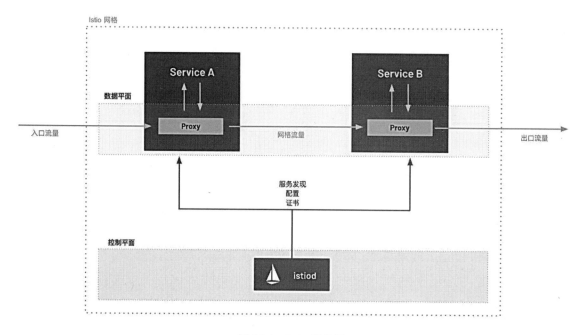

图 15-1　Istio 架构图

15.2　Istio核心组件

部署完Istio后，我们可以在Istio命名空间下找到Istio的各个组件，本节依次介绍这几个组件的功能和工作方式。

15.2.1　Istio-Pilot

Istio-Pilot是Istio服务网格架构中的关键组件，主要负责服务发现、流量管理以及配置的分发。Pilot从Kubernetes或其他注册中心获取服务信息，完成服务发现过程。它将服务信息转换为Istio能够理解的服务模型，包括服务实例的转换和服务模型的转换。此外，Pilot支持用户创建服务之间的流量转发及路由规则，并配置故障恢复策略，如超时、重试及熔断，为智能路由和弹性（如超时、重试、熔断等）提供流量管理功能。

15.2.2　Istio-Telemetry

Istio-Telemetry是Istio服务网格中的一个关键组件，主要负责遥测数据的收集和处理，主要功能是从Envoy代理收集遥测数据，包括指标、访问日志和链路追踪等。这些数据对于监控服务网格的性能、诊断问题以及优化服务至关重要。当Envoy代理处理微服务之间的网络通信时，它会将相关的遥测数据（如请求持续时间、响应码、请求大小等）发送到Istio-Telemetry。Istio-Telemetry通过适配器机制与各种后端基础设施（如Prometheus、Zipkin、Stackdriver等）集成，以便将收集到的数据发送到相应的监控系统或日志平台。还可以对收集到的数据进行预处理，如过滤、聚合和转换，以满足不同监控系统的要求。它还支持灵活的配置，允许用户根据需求自定义遥测数据的收集和处理方式。

15.2.3 Istio-Policy

Istio-Policy是Istio中的一个Mixer服务组件，它专门用于处理策略执行相关的请求。在Istio的架构中，Mixer组件通过两大核心功能——check和report，来实现对服务间访问的控制和遥测数据的收集。Istio-Policy具体负责接收来自Envoy Sidecar的check请求，并根据配置执行相应的策略检查（如授权、配额、黑白名单等），以决定是否允许请求继续访问目标服务。

Istio-Policy的工作流程如下。

01 请求拦截：当服务间的请求被Envoy Sidecar拦截时，Envoy会向Istio-Policy发送一个check请求，以检查该请求是否符合预定义的策略。

02 策略检查：Istio-Policy接收到check请求后，会根据配置文件中定义的策略规则，调用相应的适配器（Adapter）来执行具体的策略检查。这些适配器可以处理不同的策略类型，如授权、配额、黑白名单等。

03 响应决策：策略检查完成后，Istio-policy会向Envoy发送一个响应，指示该请求是被允许还是被拒绝。如果请求被允许，Envoy将继续将请求转发给目标服务；如果请求被拒绝，Envoy将终止请求并返回相应的错误响应。

15.2.4 Istio-Citadel

Istio-Citadel是Istio控制平面中的一个组件，用于管理密钥和证书，确保数据平面各服务之间的通信安全。在Istio 1.5版本之后，Citadel被整合到Istiod中，不再作为独立进程运行。

Istio-Citadel主要包括CA Server、SDS Server、Secret Controller和Monitor四个模块。

1. CA Server

（1）作为一个gRPC服务器，负责签发证书。

（2）启动时会注册两个gRPC服务：CreateCertificate（用于签发证书）和HandleCSR（用于处理CSR请求，但该方法已被CreateCertificate替代）。

（3）管理着4个证书：CA证书（ca-cert.pem）、CA私钥（ca-key.pem）、CA证书链（ca-chain.pem）和根证书（root-cert.pem）。

2. SDS Server

（1）安全发现服务（Secret Discovery Service），负责在运行时动态获取证书私钥。

（2）Envoy通过SDS动态获取证书私钥，无须重启即可更新密钥/证书对。

3. Secret Controller

监听istio.io/key-and-cert类型的Secret资源，周期性地检查证书是否过期，并更新证书。

4. Monitor

负责注册Exporter等工作，为Prometheus提供性能和指标分析。

15.2.5　Istio-Sidecar-Injector

Istio-Sidecar-Injector是一个Kubernetes的Mutating Admission Webhook，它监听Kubernetes API Server的Pod创建事件。当一个新的Pod被创建时，如果满足注入条件（例如，Pod所在的命名空间被标记为istio-injection=enabled），Istio-sidecar-injector会自动向该Pod的定义中添加Sidecar容器的配置。

Sidecar容器的作用是注入Pod中的Sidecar容器（通常是基于Envoy的Istio-Proxy容器）负责拦截和转发Pod的所有进出流量。这允许Istio控制平面管理和监控服务网格中的流量，实现路由、负载均衡、安全认证、遥测数据收集等功能。

15.2.6　Istio-Proxy

Istio-Proxy是一个基于Envoy代理的Sidecar容器，它在Istio服务网格中扮演着至关重要的角色。Envoy是一个高性能的、可扩展的、开源的代理，专为云原生应用程序设计。Istio-Proxy利用了Envoy的强大功能，包括服务发现、负载均衡、故障恢复、度量收集和安全性等，为服务网格中的服务提供全面的流量管理能力。

Istio-Proxy的工作原理主要基于以下几个步骤。

01 流量拦截：Istio-Proxy作为Sidecar容器与应用程序部署在一起，能够拦截Pod的所有进出流量。无论是进入Pod的入站流量，还是离开Pod的出站流量，都会首先经过Istio-Proxy。

02 流量处理：Istio-Proxy根据Istio控制平面（Control Plane）下发的配置信息（如路由规则、负载均衡策略等），对拦截到的流量进行处理。例如，它可以将流量路由到特定的服务实例，或者根据负载均衡算法将流量分配给不同的服务实例。

03 流量转发：处理完流量后，Istio-Proxy将流量转发给目标服务实例或客户端。同时，Istio-Proxy还会收集与流量相关的各种指标数据（如请求延迟、成功率等），并将这些数据发送给Istio Mixer（在Istio 1.5及更高版本中，Mixer的策略控制和遥测收集功能已被集成到Envoy中）。

15.2.7 Istio-Ingress-Gateway

Istio-Ingress-Gateway是一个基于Envoy的代理，它作为服务网格的入口点，允许外部流量通过HTTP/HTTPS、TCP等协议访问网格内部的服务。它提供了负载均衡、TLS终止、流量控制、安全认证等功能，是服务网格与外部世界之间的桥梁。

Istio-Ingress-Gateway的工作原理主要包括以下几个步骤。

01 接收外部流量：外部流量首先到达Istio Ingress Gateway。这些流量可以是通过LoadBalancer、NodePort或其他方式进入的。

02 流量处理：Istio Ingress Gateway根据配置的Gateway资源和VirtualService资源对流量进行处理。Gateway资源定义了要暴露的端口、协议、域名等信息，而VirtualService资源则定义了流量路由规则。

03 转发流量：处理完流量后，Istio Ingress Gateway将流量转发到网格内部相应的服务实例。这一过程可能涉及负载均衡、TLS加密、身份验证等多个环节。

15.2.8 Istio-Envoy

Istio的数据面由Envoy代理组成，Envoy是Istio数据面的核心部分，Istio数据面的绝大部分功能是由Envoy提供的。

Envoy是一款采用C++开发，面向云原生的可以独立运行的高性能服务代理，由Lyft开源，后来贡献给了CNCF。Envoy可以简单类比为Nginx，它们都是服务代理组件，但Envoy相比Nginx，除拥有更高的性能外，还拥有更多的特性，包括高级的服务治理功能（重试、断路器、全局限流、流量镜像等）、能够动态管理配置的API、深度的可观测能力。正是由于这些特性，Envoy成为云原生领域服务代理的明星组件。

Envoy和Istio并不是强绑定关系：一方面，Envoy并不局限于Istio，除在Istio数据面中应用外，以Envoy为基础的项目还有很多，比如K8s的Ingress Controller：Ambassador、Contour、EnRoute及Gloo等都是基于Envoy的Ingress控制器；另一方面，Istio也不局限于Envoy，Envoy是Istio数据面的默认选择，但不是唯一选择，Istio数据面还可以选择蚂蚁集团开源的MOSN、Nginx提供的nginMesh等。

15.3 部署Istio

本节介绍如何使用Istioctl工具在Kubernetes集群中部署Istio。

15.3.1 Istioctl 的安装

Istioctl是Istio的一个命令行工具，用于与Istio控制平面进行交互和管理。它提供了一系列命令，可以帮助用户配置路由规则、查看网格拓扑、进行故障排除等操作。要使用该工具，首先要安装它，其安装步骤如下：

01 下载Istio软件包：

```
[root@k8s-master ~]# wget https://github.com/istio/istio/releases/download/1.17.8/istio-1.17.8-linux-amd64.tar.gz
```

02 解压Istio软件包：

```
[root@k8s-master ~]# tar -xf istio-1.17.8-linux-amd64.tar.gz
```

03 将可执行文件复制到/usr/local/bin/目录下：

```
[root@k8s-master ~]# cp istio-1.17.8/bin/istioctl /usr/local/bin/
```

04 查看Istio的目录结构：

```
[root@k8s-master ~]# cd istio-1.17.8
[root@k8s-master istio-1.17.8]# ll
总用量 28
drwxr-x---  2 root root    22 7月  12 09:11 bin
-rw-r--r--  1 root root 11357 7月  12 09:11 LICENSE
drwxr-xr-x  4 root root    36 7月  12 09:11 manifests
-rw-r-----  1 root root   956 7月  12 09:11 manifest.yaml
-rw-r--r--  1 root root  6759 7月  12 09:11 README.md
drwxr-xr-x 26 root root  4096 7月  12 09:11 samples
drwxr-xr-x  3 root root    57 7月  12 09:11 tools
```

05 查看samples样品目录，官方给出了很多案例，可以运行这些案例：

```
[root@k8s-master istio-1.17.8]# ll samples/
总用量 16
drwxr-xr-x 3 root root  134 7月  12 09:11 addons
drwxr-xr-x 6 root root  122 7月  12 09:11 ambient-argo
drwxr-xr-x 7 root root  158 7月  12 09:11 bookinfo
drwxr-xr-x 2 root root   23 7月  12 09:11 builder
drwxr-xr-x 2 root root 4096 7月  12 09:11 certs
drwxr-xr-x 3 root root   22 7月  12 09:11 cicd
drwxr-xr-x 2 root root   76 7月  12 09:11 custom-bootstrap
drwxr-xr-x 4 root root   98 7月  12 09:11 extauthz
drwxr-xr-x 2 root root   78 7月  12 09:11 external
drwxr-xr-x 2 root root   45 7月  12 09:11 grpc-echo
drwxr-xr-x 2 root root   71 7月  12 09:11 health-check
drwxr-xr-x 4 root root  184 7月  12 09:11 helloworld
drwxr-xr-x 4 root root  166 7月  12 09:11 httpbin
drwxr-xr-x 4 root root   56 7月  12 09:11 jwt-server
drwxr-xr-x 2 root root   43 7月  12 09:11 kind-lb
drwxr-xr-x 2 root root  140 7月  12 09:11 multicluster
drwxr-xr-x 5 root root   61 7月  12 09:11 open-telemetry
drwxr-xr-x 2 root root 4096 7月  12 09:11 operator
drwxr-xr-x 2 root root   74 7月  12 09:11 ratelimit
-rw-r--r-- 1 root root   98 7月  12 09:11 README.md
drwxr-xr-x 4 root root   30 7月  12 09:11 security
```

```
drwxr-xr-x 2 root root   86 7月 12 09:11 sleep
drwxr-xr-x 4 root root 4096 7月 12 09:11 tcp-echo
drwxr-xr-x 3 root root   46 7月 12 09:11 wasm_modules
drwxr-xr-x 2 root root   57 7月 12 09:11 websockets
```

15.3.2 Istioctl 安装 Istio

安装好Istioctl工具后，我们就可以使用该工具来部署Istio了。具体操作步骤如下：

01 使用demo配置文件通过Istioctl工具自动安装Istio到Kubernetes集群中：

```
[root@k8s-master istio-1.17.8]# istioctl install --set profile=demo -y
✔ Istio core installed
✔ Istiod installed
✔ Ingress gateways installed
✔ Egress gateways installed
✔ Installation complete
Making this installation the default for injection and validation.

Thank you for installing Istio 1.17. Please take a few minutes to tell us about your
install/upgrade experience! https://forms.gle/hMHGiwZHPU7UQRWe9
```

Istio提供以下几种内置配置文件。

- default：根据IstioOperatorAPI的默认设置启动组件，建议用于生产部署和Multicluster Mesh中的Primary Cluster。用户可以运行istioctl profile dump命令来查看默认设置。
- demo：这一配置具有适度的资源需求，旨在展示Istio的功能。它适合运行Bookinfo应用程序和相关任务。此配置文件启用了高级别的追踪和访问日志，因此不适合进行性能测试。
- minimal：与默认配置文件相同，但只安装了控制平面组件。它允许用户使用Separate Profile配置控制平面和数据平面组件（例如Gateway）。
- remote：配置Multicluster Mesh的Remote Cluster。
- empty：不部署任何东西，可以作为自定义配置的基本配置文件。
- preview：预览文件包含的功能都是实验性的。这是为了探索Istio的新功能，不确保稳定性、安全性和性能（使用风险需自负）。

02 查看Pod和Service信息，确认Istio环境安装成功，如图15-2所示。

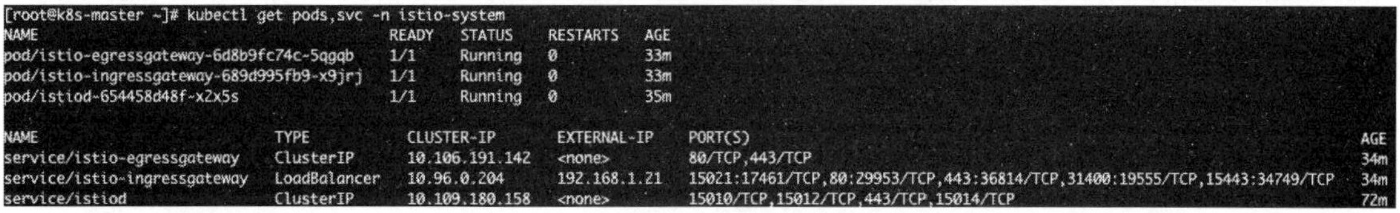

```
[root@k8s-master ~]# kubectl get pods,svc -n istio-system
NAME                                        READY   STATUS    RESTARTS   AGE
pod/istio-egressgateway-6d8b9fc74c-5qgqb    1/1     Running   0          33m
pod/istio-ingressgateway-689d995fb9-x9jrj   1/1     Running   0          33m
pod/istiod-654458d48f-x2x5s                 1/1     Running   0          35m

NAME                         TYPE           CLUSTER-IP       EXTERNAL-IP    PORT(S)                                                                      AGE
service/istio-egressgateway  ClusterIP      10.106.191.142   <none>         80/TCP,443/TCP                                                               34m
service/istio-ingressgateway LoadBalancer   10.96.0.204      192.168.1.21   15021:17461/TCP,80:29953/TCP,443:36814/TCP,31400:19555/TCP,15443:34749/TCP   34m
service/istiod               ClusterIP      10.109.180.158   <none>         15010/TCP,15012/TCP,443/TCP,15014/TCP                                        72m
```

图 15-2 Istio 安装成功后的 Pod 资源状态效果

03 部署Kiali仪表板以及Prometheus、Grafana和Jaeger：

```
[root@k8s-master istio-1.17.8]# kubectl apply -f samples/addons/
serviceaccount/grafana created
configmap/grafana created
service/grafana created
deployment.apps/grafana created
configmap/istio-grafana-dashboards created
configmap/istio-services-grafana-dashboards created
deployment.apps/jaeger created
service/tracing created
service/zipkin created
service/jaeger-collector created
serviceaccount/kiali created
configmap/kiali created
clusterrole.rbac.authorization.k8s.io/kiali-viewer created
clusterrole.rbac.authorization.k8s.io/kiali created
clusterrolebinding.rbac.authorization.k8s.io/kiali created
role.rbac.authorization.k8s.io/kiali-controlplane created
rolebinding.rbac.authorization.k8s.io/kiali-controlplane created
service/kiali created
deployment.apps/kiali created
serviceaccount/prometheus created
configmap/prometheus created
clusterrole.rbac.authorization.k8s.io/prometheus created
clusterrolebinding.rbac.authorization.k8s.io/prometheus created
service/prometheus created
deployment.apps/prometheus created
```

04 查看Pod和Service资源信息，如图15-3所示。

```
[root@k8s-master istio-1.17.8]# kubectl get pods -n istio-system
NAME                                    READY   STATUS    RESTARTS   AGE
grafana-6d674c7b55-k28ss                1/1     Running   0          11m
istio-egressgateway-6d8b9fc74c-5qgqb    1/1     Running   0          47h
istio-ingressgateway-689d995fb9-x9jrj   1/1     Running   0          47h
istiod-654458d48f-x2x5s                 1/1     Running   0          47h
jaeger-7ff9f69fff-8kzpc                 1/1     Running   0          10m
kiali-576f9965dd-m7nnl                  1/1     Running   0          10m
prometheus-5d7cd5bc-4lkpf               2/2     Running   0          4m35s
[root@k8s-master istio-1.17.8]# kubectl get svc -n istio-system
NAME                   TYPE           CLUSTER-IP       EXTERNAL-IP    PORT(S)                                                                      AGE
grafana                ClusterIP      10.110.171.216   <none>         3000/TCP                                                                     41m
istio-egressgateway    ClusterIP      10.106.191.142   <none>         80/TCP,443/TCP                                                               2d
istio-ingressgateway   LoadBalancer   10.96.0.204      192.168.1.21   15021:17461/TCP,80:29953/TCP,443:36814/TCP,31400:19555/TCP,15443:34749/TCP   2d
istiod                 ClusterIP      10.109.180.158   <none>         15010/TCP,15012/TCP,443/TCP,15014/TCP                                        2d
jaeger-collector       ClusterIP      10.102.118.104   <none>         14268/TCP,14250/TCP,9411/TCP                                                 41m
kiali                  ClusterIP      10.108.185.67    <none>         20001/TCP,9090/TCP                                                           41m
prometheus             ClusterIP      10.107.74.149    <none>         9090/TCP                                                                     41m
tracing                ClusterIP      10.103.196.61    <none>         80/TCP,16685/TCP                                                             41m
zipkin                 ClusterIP      10.103.6.211     <none>         9411/TCP                                                                     41m
```

图15-3　Istio 的 Pod 和 Service 资源信息

05 再次查看版本号，可以看出客户端版本和控制平面以及数据平面的版本：

```
[root@k8s-master ~]# istioctl version
client version: 1.17.8
control plane version: 1.17.8
data plane version: 1.17.8 (2 proxies)
```

06 创建Kiali的Ingress资源清单：

```
[root@k8s-master ~]# vim kiali-ingress.yaml
apiVersion: networking.k8s.io/v1
kind: Ingress
metadata:
  name: kiali
  namespace: istio-system
spec:
  ingressClassName: nginx
  rules:
  - host: kiali.deanit.cn
    http:
      paths:
      - backend:
          service:
            name: kiali
            port:
              number: 20001
        path: /
        pathType: Prefix
```

07 应用资源：

```
[root@k8s-master ~]# kubectl apply -f kiali-ingress.yaml
ingress.networking.k8s.io/kiali created
```

08 查看istio-system命名空间下Ingress的资源信息：

```
[root@k8s-master ~]# kubectl get ing -n istio-system
NAME    CLASS   HOSTS             ADDRESS          PORTS   AGE
kiali   nginx   kiali.deanit.cn   10.110.206.243   80      118s
```

09 由于我们创建了Ingress资源，因此可以通过域名来访问Kiali服务，各客户端配置如下：

```
# Linux和macOS客户端配置
$ vim /etc/hosts
192.168.1.11 kiali.deanit.cn

# Windows客户端配置
# hosts文件位于路径：C:\Windows\System32\drivers\etc\hosts
192.168.1.11 kiali.deanit.cn
```

> **注意** 由于hosts文件属于系统文件，因此在修改之前需要获取管理员权限。

10 通过域名访问后，Istio首页如图15-4所示。

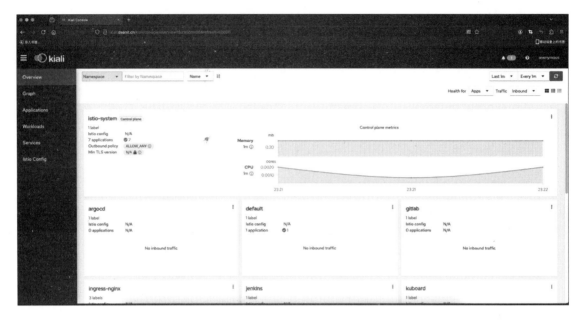

图 15-4　Istio 首页

至此，我们成功地安装了Istio，接下来介绍 Istio的使用。

15.4　Sidecar边车容器注入

Sidecar边车容器注入，特别是在Istio和Kubernetes环境中，是一个关键的过程，它允许为每个Pod自动注入一个或多个辅助容器（Sidecar），以实现流量管理、安全、监控等功能。

15.4.1　Sidecar 手动注入

手动注入是针对一个具体的资源对象的。

工作原理：通过改写YAML文件植入Istio-proxy容器，以及init-container用于修改iptables规则。

配置方法：

```
istioctl kube-inject [app.yaml] > [inject.yaml] | kubectl apply -f -
```

01 在k8s-master节点创建一个资源清单：

```
[root@k8s-master ~]# vim nginx-deployment.yaml
apiVersion: v1
kind: Namespace
metadata:
  name: demo
---
```

```yaml
apiVersion: apps/v1
kind: Deployment
metadata:
  labels:
    app: nginx-deployment
  name: nginx-deployment
  namespace: demo
spec:
  replicas: 2
  selector:
    matchLabels:
      app: nginx-deployment
  strategy:
    rollingUpdate:
      maxSurge: 25%
      maxUnavailable: 25%
    type: RollingUpdate
  template:
    metadata:
      labels:
        app: nginx-deployment
    spec:
      containers:
        - image: 'nginx:latest'
          name: nginx-deployment
```

02 运行手动注入命令生成新的资源定义清单：

```
[root@k8s-master ~]# istioctl kube-inject -f nginx-deployment.yaml > nginx-deployment-inject.yaml
```

参数解析：

- istioctl kube-inject -f nginx-deployment.yaml：读取nginx-deployment.yaml文件，向其中注入Istio的sidecar容器配置。
- >：将命令的输出重定向到文件。
- nginx-deployment-inject.yaml：指定输出文件的名称，这里是将注入后的内容保存到nginx-deployment-inject.yaml文件中。

此时会发现nginx-deployment-inject.yaml文件新增了proxy容器和init容器：

```yaml
        image: docker.io/istio/proxyv2:1.17.8
        name: istio-proxy
        ports:
        - containerPort: 15090
          name: http-envoy-prom
          protocol: TCP
        readinessProbe:
          failureThreshold: 30
```

```yaml
        httpGet:
          path: /healthz/ready
          port: 15021
        initialDelaySeconds: 1
        periodSeconds: 2
        timeoutSeconds: 3
      resources:
        limits:
          cpu: "2"
          memory: 1Gi
        requests:
          cpu: 10m
          memory: 40Mi
      securityContext:
        allowPrivilegeEscalation: false
        capabilities:
          drop:
          - ALL
        privileged: false
        readOnlyRootFilesystem: true
        runAsGroup: 1337
        runAsNonRoot: true
        runAsUser: 1337
      volumeMounts:
      - mountPath: /var/run/secrets/istio
        name: istiod-ca-cert
      - mountPath: /var/lib/istio/data
        name: istio-data
      - mountPath: /etc/istio/proxy
        name: istio-envoy
      - mountPath: /var/run/secrets/tokens
        name: istio-token
      - mountPath: /etc/istio/pod
        name: istio-podinfo
  initContainers:
  - args:
    - istio-iptables
    - -p
    - "15001"
    - -z
    - "15006"
    - -u
    - "1337"
    - -m
    - REDIRECT
    - -i
    - '*'
    - -x
    - ""
    - -b
```

```
        - '*'
        - -d
        - 15090,15021,15020
```

03 应用资源：

```
[root@k8s-master ~]# kubectl apply -f nginx-deployment-inject.yaml
namespace/demo created
deployment.apps/nginx-deployment created
```

04 查看Pod资源信息，可以看到每个Pod中有两个容器，分别为nginx和proxy：

```
[root@k8s-master ~]# kubectl get pods -n demo
NAME                                READY   STATUS    RESTARTS   AGE
nginx-deployment-7d9d5c8857-bnmvj   2/2     Running   0          104s
nginx-deployment-7d9d5c8857-fzj8s   2/2     Running   0          104s
```

15.4.2 Sidecar 自动注入

自动注入是针对命名空间内的所有资源对象的。

工作原理：通过Webhook机制监听apiserver提交的请求，从而自动修改对应的资源对象。

配置方法：

```
kubectl label namespace <namespace> istio-injection=enabled
```

01 设置default命名空间，开启自动注入Envoy Sidecar：

```
--- 给命名空间添加标签，指示Istio在部署应用时，自动注入Envoy边车代理
[root@k8s-master ~]# kubectl label namespace default istio-injection=enabled
namespace/default labeled
```

02 查看命名空间istio-injection标签，可以看到default命名空间标签的值为enable：

```
[root@k8s-master ~]# kubectl get namespace -L istio-injection
NAME                STATUS   AGE    ISTIO-INJECTION
argocd              Active   11d
default             Active   85d    enabled
demo                Active   17m
gitlab              Active   14d
ingress-nginx       Active   48d
istio-system        Active   3d8h
jenkins             Active   13d
kube-node-lease     Active   85d
kube-public         Active   85d
kube-system         Active   85d
kuboard             Active   85d
metallb-system      Active   10d
monitoring          Active   8d
myapp-prod          Active   11d
myapp-uat           Active   11d
```

```
nfs-client-provisioner   Active   36d
```

运维笔记：原有运行的Pod，不会被赋予Sidecar边车容器，如果这个Pod被重启，那么会被注入Sidecar边车容器。

03 重启default命名空间下名为lb-nginx的资源：

```
[root@k8s-master ~]# kubectl rollout restart deployment/lb-nginx
deployment.apps/lb-nginx restarted
```

04 持续观察default命名空间下名为lb-nginx的Pod状态信息：

```
--- 创建新的Pod，并且里面有两个容器
[root@k8s-master ~]# kubectl get pods
NAME                            READY   STATUS     RESTARTS   AGE
lb-nginx-5d9cdb7848-skkb4       1/1     Running    0          10d
lb-nginx-d8df75bbd-g9qx2        0/2     Init:0/1   0          14s

--- 新的Pod启动完成后，旧的Pod已经被删除
[root@k8s-master ~]# kubectl get pods
NAME                            READY   STATUS    RESTARTS   AGE
lb-nginx-d8df75bbd-g9qx2        2/2     Running   0          48s
```

15.4.3 Sidecar 取消自动注入

在default命名空间开启自动注入Envoy Sidecar：

```
[root@k8s-master ~]# kubectl label namespaces default istio-injection-
namespace/default unlabeled
```

重启default命名空间下名为lb-nginx的资源并持续观察Pod状态：

```
--- 重启default命名空间下名为lb-nginx的资源
[root@k8s-master ~]# kubectl rollout restart deployment/lb-nginx
deployment.apps/lb-nginx restarted

--- 持续观察default命名空间下名为lb-nginx的Pod状态信息
[root@k8s-master ~]# kubectl get pods
NAME                            READY   STATUS              RESTARTS   AGE
lb-nginx-6b76678b76-6hp98       0/1     ContainerCreating   0          1s
lb-nginx-d8df75bbd-g9qx2        2/2     Running             0          6m18s

--- 新的Pod创建完成后，旧的Pod已经被删除，并且Pod中只有一个容器
[root@k8s-master ~]# kubectl get pods
NAME                            READY   STATUS    RESTARTS   AGE
lb-nginx-6b76678b76-6hp98       1/1     Running   0          57s
```

15.5　4种配置资源概念详解

Istio是一个开源的服务网格，它主要用于简化微服务架构中的服务间通信、提供强大的监控能力以及加强服务的安全管理。Istio中包含4种核心的流量管理配置资源，它们分别是VirtualService、DestinationRule、ServiceEntry以及Gateway。本节主要介绍这4种配置资源及其应用场景。

通过查看请求链路图，可以更直观地了解请求从外部到集群服务内的整个请求过程，如图15-5所示。

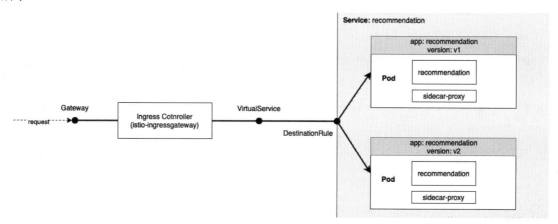

图 15-5　请求链路图

15.5.1　VirtualService

VirtualService用于在Istio中定义路由规则，控制流量如何路由到服务上。它主要用于声明一个后端服务及其路由信息，并可以实现流量管控（如镜像流量、故障注入、跨域配置等）。通过VirtualService可以定义复杂的路由逻辑，将流量路由到不同的服务版本或基于特定条件（如HTTP路径、请求头）进行路由。

VirtualService的应用场景如下。

- 灰度发布：将部分流量路由到新版本的服务上，以测试新功能的稳定性和性能。
- 流量切分：基于不同的条件（如用户地理位置、请求时间等）将流量路由到不同的服务实例。
- 跨域配置：配置HTTP请求的跨域资源共享（Cross-Origin Resource Sharing，CORS）策略。

15.5.2　DestinationRule

DestinationRule在VirtualService路由生效后，配置应用与请求策略。它定义了对应目标主机的可路由子集（如具有特定版本的实例），并为这些子集指定了负载均衡策略、断路器、TLS配置等。DestinationRule还允许对目标服务进行更细粒度的控制，如健康检查、连接池设置等。

DestinationRule的应用场景如下。

- 负载均衡：配置服务的负载均衡策略，如轮询、一致性哈希等。
- 断路器：设置服务的断路器策略，防止因某个服务故障而导致整个系统崩溃。
- TLS配置：为服务间的通信配置TLS加密，增强安全性。

15.5.3　ServiceEntry

ServiceEntry通常用于在Istio服务网格之外启用对服务的请求。它允许将外部服务（如数据库、第三方API等）添加到Istio的服务注册表中，以便在服务网格内部对这些服务进行路由和访问控制。ServiceEntry可以配置为网格内部或网格外部的服务，并支持双向TLS认证（对于网格内部的服务）。

ServiceEntry的应用场景如下。

- 访问外部服务：将外部数据库、第三方API等集成到Istio服务网格中，以便进行统一的管理和访问控制。
- 网格扩展：将不受管理的虚拟机或物理机上的服务加入Istio服务网格中，以实现服务的统一治理。

15.5.4　Gateway

Gateway为HTTP/TCP流量配置了一个负载均衡器，主要用于启用一个服务的Ingress流量。它定义了监听信息与入口网关的绑定关系，允许外部流量通过特定的域名和端口进入服务网格，并根据VirtualService和DestinationRule将请求路由到后端的Kubernetes Service。

Gateway的应用场景如下。

- 入口流量管理：配置外部流量进入服务网格的方式和规则，实现流量路由、负载均衡、TLS加密等功能。
- 多租户支持：在多租户环境中，为每个租户或一组相关服务部署独立的Ingress Gateway，以实现网络流量的有效隔离和定制化配置。

15.6　VirtualService关键字配置示例

通过VirtualService可以定义如何将流量路由到不同的服务版本、设置请求的超时时间、配置重试策略以及注入故障以进行测试。本节介绍如何使用VirtualService关键字来配置这些功能的示例。

15.6.1　使用 weight 关键字拆分流量

根据不同的版本对服务流量进行拆分是常用的功能。在Istio中，服务版本依靠标签进行区分，可以定义不同种类的标签（如版本号、平台），对流量以不同的维度进行灵活的分配。拆分流量使用weight关键字来设置。

通过以下配置，把75%的流量分配给v1版本的reviews服务，25%的流量分配给v2版本的reviews服务：

```yaml
apiVersion: networking.istio.io/v1beta1
kind: VirtualService
metadata:
  name: reviews
spec:
  hosts:
    - reviews
  http:
  - route:
    - destination:
        host: reviews
        subset: v1
      weight: 75
    - destination:
        host: reviews
        subset: v2
      weight: 25
```

上面的配置中出现了subset（子集）关键字。subset其实就是特定版本的标签，它和标签的映射关系定义在DestinationRule中。例如在subset中设置标签为version: v1，代表只有附带这个标签的Pod才能接受流量。Istio强制要求Pod设置带有version的标签，以此来实现流量控制和指标收集等功能。

比如下面的例子定义了两个子集，分别对应v1、v2两个标签：

```yaml
subsets:                # 定义了包含两个标签的子集
- name: v1
  labels:
    version: v1         # 真实的Pod标签
- name: v2
  labels:
    version: v2
```

15.6.2 使用 timeout 关键字设置请求超时时间

对访问ratings服务的请求设置10s超时：

```yaml
apiVersion: networking.istio.io/v1beta1
kind: VirtualService
metadata:
  name: ratings
spec:
  hosts:
    - ratings
  http:
  - route:
    - destination:
```

```
      host: ratings
        subset: v1
  timeout: 10s
```

15.6.3 使用 retries 关键字设置重试

设置表示最多重试3次,每次的超时时间为2s:

```
apiVersion: networking.istio.io/v1beta1
kind: VirtualService
metadata:
  name: ratings
spec:
  hosts:
    - ratings
  http:
  - route:
    - destination:
        host: ratings
        subset: v1
    retries:
      attempts: 3
      perTryTimeout: 2s
```

15.6.4 使用 fault 关键字设置故障注入

在微服务应用中,时常会出现服务网络延迟等问题,Istio中提供了配置可以模拟这些错误,这样就可以测试在真实环境中发生问题时,上下游的服务是否会受到影响,并做出相应的调整。这就是故障注入功能。

注入一个延迟故障,可以使得ratings服务10%的响应出现5s的延迟。除延迟外,还可以设置终止或者返回HTTP故障码:

```
apiVersion: networking.istio.io/v1beta1
kind: VirtualService
metadata:
  name: ratings
spec:
  hosts:
    - ratings
  http:
  - fault:
      delay:
        percent: 10
        fixedDelay: 5s
    route:
    - destination:
        host: ratings
        subset: v1
```

15.6.5 VirtualService 资源清单详解

```
apiVersion: networking.istio.io/v1beta1
kind: VirtualService
metadata:
  name: proxy
spec:
# 用于定义哪个路由和虚拟主机有关系，所以需要指定hosts,此hosts必须和GW中的hosts保持一致或者包
含关系
  hosts:
  - "test.deanit.cn"                  # 对应gateways/proxy-gateway
  # (和网关相关联)
  # gateways用于指定该 vs 是定义在Ingress Gateway的接收入栈流量，并指定GW名称
  gateways:
  - istio-system/proxy-gateway        # 相关定义仅应用于Ingress Gateway上
  #- mesh
  # HTTP七层路由
  http:
  #  路由策略名称
  - name: default
    # 路由目标
    route:
    - destination:
        # proxy cluster 是被自动生成的,因为集群内部有一个同名的Service,而且此集群在ingess
gateway上本身存在
        # 内部集群Service名称，但是流量不会直接发给Service,而是发给由Service组成的集群(这里的
七层调度流量不再经由Service)
        host: proxy      # proxy是要对demoapp 发起请求的
    # 超时时长设置为1s,如果服务端要处理1s以上才会回复响应，则会立即响应客户端为超时
    timeout: 1s
    # 重试策略
    retries:
      # 重试次数
      attempts: 5
      # 重试操作的超时时长,如果重试时超过1s,则会响应客户端为超时
      perTryTimeout: 1s
      # 重试的条件:对于后端服务端返回的5xx系列的响应状态码,会转到retries 重试策略
      retryOn: 5xx,connect-failure,refused-stream
```

15.7 Istio流量治理

　　在微服务架构中，流量治理是确保系统稳定性和高效性的关键。Istio作为一个服务网格，提供了丰富的流量管理功能，包括请求头处理、流量镜像、重写、重定向以及蓝绿部署和金丝雀发布等。本节将详细介绍如何使用Istio进行流量治理，帮助读者了解如何通过配置不同的策略来满足特定的业务需求。

15.7.1 请求头 httpHeaderName

请求标头包含X-User标头,如果X-User标头的值相同,则将请求调度到同一个后端Pod;如果值不同,则重新调度。

01 创建资源清单:

```yaml
[root@k8s-master ~]# vim httpHeaderName.yaml
---
# 暴露网关
apiVersion: networking.istio.io/v1beta1
kind: Gateway
metadata:
  name: nginx-gateway
spec:
  selector:
    app: istio-ingressgateway
  servers:
    - port:
        number: 80
        name: http
        protocol: HTTP
      hosts:
        - "httpheadername.deanit.cn"
---
# 版本 1 的服务
apiVersion: apps/v1
kind: Deployment
metadata:
  labels:
    app: nginx
  name: nginx-v1-deployment
spec:
  replicas: 2
  selector:
    matchLabels:
      app: nginx
      version: v1
  template:
    metadata:
      labels:
        app: nginx
        version: v1
    spec:
      containers:
        - image: registry.cn-hangzhou.aliyuncs.com/deanmr/dean:myappv1
          name: nginx-deployment
---
# svc,指向两个版本的服务
```

```yaml
apiVersion: v1
kind: Service
metadata:
  name: nginx-service
spec:
  ports:
    - name: http
      port: 80
      protocol: TCP
      targetPort: 80
  selector:
    app: nginx
  type: ClusterIP
---
# 目标规则
apiVersion: networking.istio.io/v1beta1
kind: DestinationRule
metadata:
  name: nginx-dr
spec:
  host: nginx-service
  # 全局负载均衡策略
  trafficPolicy:
    # 负载均衡策略
    loadBalancer:
      # 加权最少连接
      simple: LEAST_CONN
  subsets:
    - name: v1
      labels:
        version: v1
      ############定义策略############
      trafficPolicy:
        # 负载均衡策略
        loadBalancer:
          # 一致性哈希
          consistentHash:
            # 当请求标头有X-User时，如果X-User标头值一样，则将调度到同一个后端Pod，如果不一样，则被重新调度
            httpHeaderName: X-User
---
# 虚拟服务
apiVersion: networking.istio.io/v1beta1
kind: VirtualService
metadata:
  name: nginx-vs
spec:
  hosts:
    - "*"
  gateways:
```

```
        - nginx-gateway
  # HTTP七层路由
  http:
      # 路由目标，正常流量全部发往 nginx-service v1 子集
      - route:
        - destination:
            host: nginx-service
            subset: v1
```

02 应用资源：

```
[root@k8s-master ~]# kubectl apply -f httpHeaderName.yaml
gateway.networking.istio.io/nginx-gateway created
deployment.apps/nginx-v1-deployment created
service/nginx-service created
destinationrule.networking.istio.io/nginx-dr created
virtualservice.networking.istio.io/nginx-vs created
```

03 查看资源状态信息：

```
[root@k8s-master ~]# kubectl get pods,svc,gateway,vs,dr
NAME                                              READY   STATUS    RESTARTS   AGE
pod/fortio-deploy-65564fc64d-fnrhr                2/2     Running   0          3h52m
pod/lb-nginx-6b76678b76-mrbmt                     2/2     Running   0          4h47m
pod/nginx-v1-deployment-5fd9c5d9f6-6rzsc          2/2     Running   0          3m42s
pod/nginx-v1-deployment-5fd9c5d9f6-h4j6k          2/2     Running   0          3m42s

NAME                    TYPE           CLUSTER-IP       EXTERNAL-IP      PORT(S)        AGE
service/fortio          ClusterIP      10.99.147.122    <none>           8080/TCP       4h49m
service/kubernetes      ClusterIP      10.96.0.1        <none>           443/TCP        86d
service/nginx-service   ClusterIP      10.99.144.187    <none>           80/TCP         3m42s
service/lb-nginx        LoadBalancer   10.108.218.198   192.168.1.20     80:35139/TCP   11d
service/myapp-svcname   ExternalName   <none>           myapp-svc.dean.svc.cluster.local   <none>   54d

NAME                                          AGE
gateway.networking.istio.io/nginx-gateway     3m43s

NAME                                              GATEWAYS           HOSTS   AGE
virtualservice.networking.istio.io/nginx-vs       ["nginx-gateway"]  ["*"]   3m43s

NAME                                              HOST            AGE
destinationrule.networking.istio.io/nginx-dr      nginx-service   3m42s
```

04 查看istio-ingressgateway的LoadBalancer分配的IP地址，如图15-6所示。

05 在k8s-master节点配置hosts进行域名解析，将域名解析到istio-ingressgateway的LoadBalancer分配的IP地址上：

```
[root@k8s-master ~]# echo "192.168.1.21 httpheadername.deanit.cn">>/etc/hosts
```

```
[root@k8s-master ~]# kubectl get svc -n istio-system
NAME                    TYPE            CLUSTER-IP        EXTERNAL-IP     PORT(S)                                                                        AGE
grafana                 ClusterIP       10.110.171.216    <none>          3000/TCP                                                                       2d5h
istio-egressgateway     ClusterIP       10.106.191.142    <none>          80/TCP,443/TCP                                                                 4d5h
istio-ingressgateway    LoadBalancer    10.96.0.204       192.168.1.21    15021:17461/TCP,80:29953/TCP,443:36814/TCP,31400:19555/TCP,15443:34749/TCP     4d5h
istiod                  ClusterIP       10.109.180.158    <none>          15010/TCP,15012/TCP,443/TCP,15014/TCP                                          4d6h
jaeger-collector        ClusterIP       10.102.118.104    <none>          14268/TCP,14250/TCP,9411/TCP                                                   2d5h
kiali                   ClusterIP       10.108.185.67     <none>          20001/TCP,9090/TCP                                                             2d5h
prometheus              ClusterIP       10.107.74.149     <none>          9090/TCP                                                                       2d5h
tracing                 ClusterIP       10.103.196.61     <none>          80/TCP,16685/TCP                                                               2d5h
zipkin                  ClusterIP       10.103.6.211      <none>          9411/TCP                                                                       2d5h
```

图 15-6 Istio 集群的 Service 信息

06 请求测试，可以看到请求到同一个后端Pod：

```
[root@k8s-master ~]# while sleep 1; do curl -H "X-User: user" httpheadername.deanit.cn/hostname.html; done
    nginx-v1-deployment-5fd9c5d9f6-h4j6k
    nginx-v1-deployment-5fd9c5d9f6-h4j6k
    nginx-v1-deployment-5fd9c5d9f6-h4j6k
    nginx-v1-deployment-5fd9c5d9f6-h4j6k
    nginx-v1-deployment-5fd9c5d9f6-h4j6k
    nginx-v1-deployment-5fd9c5d9f6-h4j6k
    nginx-v1-deployment-5fd9c5d9f6-h4j6k
    nginx-v1-deployment-5fd9c5d9f6-h4j6k
```

07 通过查看Kiali图形化，可以看到请求的整个动态链路图，将Kiali的Service改为NodePort，然后访问任意节点IP+NodePort端口即可，如图15-7所示。

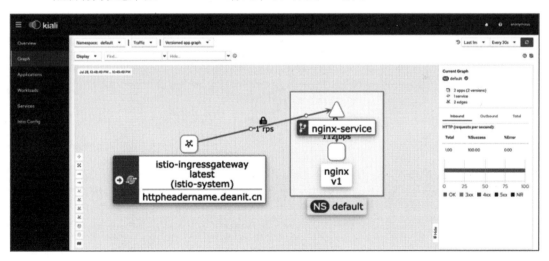

图 15-7 服务请求链路图

15.7.2 HTTP 流量镜像

例如，如果上线了新版本但对其可靠性不够有信心，可以将部分请求流量从当前使用的版本镜像引流至新版本进行测试。这样做不会对当前使用的版本产生影响，并且可以同时测试新版本在性能等方面的表现。

01 创建命名空间：

```
[root@k8s-master ~]# kubectl create ns mirror
namespace/mirror created
```

02 设置Sidecar自动注入：

```
[root@k8s-master ~]# kubectl label namespace mirror istio-injection=enabled
namespace/mirror labeled
```

03 创建资源清单：

```yaml
[root@k8s-master ~]# vim mirror.yaml
---
# 暴露网关
apiVersion: networking.istio.io/v1beta1
kind: Gateway
metadata:
  name: mirror-gateway
  namespace: mirror
spec:
  selector:
    istio: ingressgateway
  servers:
    - port:
        number: 80
        name: http
        protocol: HTTP
      hosts:
        - "mirror.deanit.cn"
---
# 版本 1 的服务
apiVersion: apps/v1
kind: Deployment
metadata:
  labels:
    app: mirror
  name: myapp-v1
  namespace: mirror
spec:
  replicas: 1
  selector:
    matchLabels:
      app: mirror
      version: v1
  template:
    metadata:
      labels:
        app: mirror
        version: v1
    spec:
```

```yaml
      containers:
        - image: registry.cn-hangzhou.aliyuncs.com/deanmr/dean:myappv1
          name: mirror
---
# 版本 2 的服务
apiVersion: apps/v1
kind: Deployment
metadata:
  labels:
    app: mirror
  name: myapp-v2
  namespace: mirror
spec:
  replicas: 1
  selector:
    matchLabels:
      app: mirror
      version: v2
  template:
    metadata:
      labels:
        app: mirror
        version: v2
    spec:
      containers:
        - image: registry.cn-hangzhou.aliyuncs.com/deanmr/dean:myappv2
          name: mirror
---
# svc,指向两个版本的服务
apiVersion: v1
kind: Service
metadata:
  name: mirror-service
  namespace: mirror
spec:
  ports:
    - name: http
      port: 80
      protocol: TCP
      targetPort: 80
  selector:
    app: mirror
  type: ClusterIP
---
# 目标规则
apiVersion: networking.istio.io/v1beta1
kind: DestinationRule
metadata:
  name: mirror-dr
  namespace: mirror
```

```yaml
spec:
  host: mirror-service
  subsets:
    - name: v1
      labels:
        version: v1
    - name: v2
      labels:
        version: v2
---
# 虚拟服务
apiVersion: networking.istio.io/v1beta1
kind: VirtualService
metadata:
  name: mirror-vs
  namespace: mirror
spec:
  hosts:
    - "mirror.deanit.cn"
  gateways:
    - mirror-gateway
  # HTTP七层路由
  http:
      # 路由目标，正常流量全部发往 mirror-service v1 子集
      - route:
          - destination:
              host: mirror-service
              subset: v1
        # 并将100%流量镜像一份到mirror-service v2子集
        mirror:
          host: mirror-service
          subset: v2
        # 旧版本，已弃用
        # mirror_percent: 100
        # 新版本配置参数
        mirrorPercentage:
          value: 100 # 流量百分比
```

04 应用资源：

```
[root@k8s-master ~]# kubectl apply -f mirror.yaml
gateway.networking.istio.io/mirror-gateway created
deployment.apps/myapp-v1 created
deployment.apps/myapp-v2 created
service/mirror-service created
destinationrule.networking.istio.io/mirror-dr created
virtualservice.networking.istio.io/mirror-vs created
```

05 查看资源状态信息：

```
[root@k8s-master ~]# kubectl get pods,svc,gateway,vs,dr -n mirror
NAME                                READY   STATUS    RESTARTS   AGE
pod/myapp-v1-c545d58bb-7lwkb        2/2     Running   0          2m56s
pod/myapp-v2-6874b479fd-8nt72       2/2     Running   0          2m56s

NAME                      TYPE        CLUSTER-IP       EXTERNAL-IP   PORT(S)   AGE
service/mirror-service    ClusterIP   10.107.226.248   <none>        80/TCP    2m54s

NAME                                            AGE
gateway.networking.istio.io/mirror-gateway      2m56s

NAME                                         GATEWAYS            HOSTS                AGE
virtualservice.networking.istio.io/mirror-vs ["mirror-gateway"]  ["mirror.deanit.cn"] 2m56s

NAME                                             HOST             AGE
destinationrule.networking.istio.io/mirror-dr    mirror-service   2m54s
```

06 在k8s-master节点配置hosts进行域名解析，将域名解析到istio-ingressgateway的LoadBalancer分配的IP地址上：

```
[root@k8s-master ~]# echo "192.168.1.21 mirror.deanit.cn">>/etc/hosts
```

07 进行请求测试：

```
[root@k8s-master ~]# while sleep 1; do curl mirror.deanit.cn/hostname.html; done
# 输出信息
myapp-v1-c545d58bb-qgfts
myapp-v1-c545d58bb-qgfts
myapp-v1-c545d58bb-qgfts
myapp-v1-c545d58bb-qgfts
myapp-v1-c545d58bb-qgfts
myapp-v1-c545d58bb-qgfts
myapp-v1-c545d58bb-qgfts
myapp-v1-c545d58bb-qgfts
myapp-v1-c545d58bb-qgfts
```

通过查看Kiali图形化，可以看到请求的整个动态链路图，将Kiali的Service改为NodePort，然后访问任意节点IP+NodePort端口即可，通过图15-8可以看出访问v1的服务，请求流量同样会发送到v2服务。

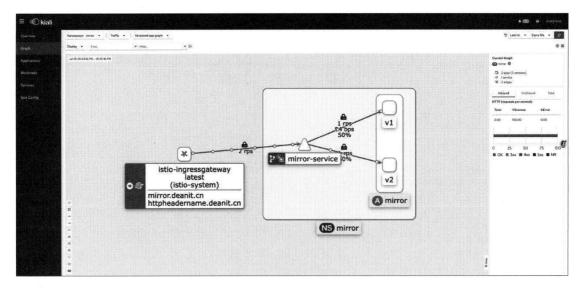

图 15-8　服务请求链路图

15.7.3　重写

在生产环境中，经常会遇到需要变更某些URL的情况，同时要求这些变更对用户请求保持透明且友好。另外，有时还需要为一些URL路径设置额外的别名。这些需求都可以通过URL重写功能简单地实现。

01 创建命名空间：

```
[root@k8s-master ~]# kubectl create ns rewrite
namespace/rewrite created
```

02 设置Sidecar自动注入：

```
[root@k8s-master ~]# kubectl label namespace rewrite istio-injection=enabled
namespace/rewrite labeled
```

03 创建资源清单：

```
[root@k8s-master ~]# vim rewrite.yaml
---
# 暴露网关
apiVersion: networking.istio.io/v1beta1
kind: Gateway
metadata:
  name: rewrite-gateway
  namespace: rewrite
spec:
  selector:
    istio: ingressgateway
```

```yaml
  servers:
    - port:
        number: 80
        name: http
        protocol: HTTP
      hosts:
        - "rewrite.deanit.cn"
---
# myapp-v1资源
apiVersion: apps/v1
kind: Deployment
metadata:
  name: rewrite-myapp-v1
  namespace: rewrite
  labels:
    server: rewrite-myapp-v1
spec:
  replicas: 1
  selector:
    matchLabels:
      server: rewrite-myapp-v1
  template:
    metadata:
      name: rewrite-myapp-v1
      labels:
        server: rewrite-myapp-v1
    spec:
      containers:
        - name: rewrite-myapp-v1
          image: registry.cn-hangzhou.aliyuncs.com/deanmr/dean:myappv1
          ports:
            - containerPort: 80
---
# myapp-v1服务
apiVersion: v1
kind: Service
metadata:
  name: rewrite-myapp-v1
  namespace: rewrite
spec:
  type: ClusterIP
  selector:
    server: rewrite-myapp-v1
  ports:
  - name: http
    port: 80
    targetPort: 80
    protocol: TCP
---
# VirtualService负责权重
```

```yaml
apiVersion: networking.istio.io/v1beta1
kind: VirtualService
metadata:
  name: rewrite-myapp-vs
  namespace: rewrite
spec:
  hosts:
  - "rewrite.deanit.cn"
  gateways:
    - rewrite-gateway
  http:
  - match:
    - uri:
        prefix: /
    rewrite:
      uri: /hostname.html
    route:
    - destination:
        host: rewrite-myapp-v1
```

04 应用资源：

```
[root@k8s-master ~]# kubectl apply -f rewrite.yaml
gateway.networking.istio.io/rewrite-gateway created
deployment.apps/rewrite-myapp-v1 created
service/rewrite-myapp-v1 created
virtualservice.networking.istio.io/rewrite-myapp-vs created
```

05 查看资源状态信息：

```
[root@k8s-master ~]# kubectl get pods,svc,gateway,vs,dr -n rewrite
NAME                                         READY   STATUS    RESTARTS   AGE
pod/rewrite-myapp-v1-d4659d574-cqc2h         2/2     Running   0          69s

NAME                           TYPE        CLUSTER-IP     EXTERNAL-IP   PORT(S)   AGE
service/rewrite-myapp-v1       ClusterIP   10.103.8.220   <none>        80/TCP    69s

NAME                                               AGE
gateway.networking.istio.io/rewrite-gateway        69s

NAME                                                   GATEWAYS             HOSTS                     AGE
virtualservice.networking.istio.io/rewrite-myapp-vs    ["rewrite-gateway"]  ["rewrite.deanit.cn"]     68s
```

06 在k8s-master节点配置hosts进行域名解析，将域名解析到istio-ingressgateway的LoadBalancer分配的IP地址上：

```
[root@k8s-master ~]# echo "192.168.1.21 rewrite.deanit.cn">>/etc/hosts
```

07 进行请求测试，可以看出访问/路径时被重写到/hostname.html：

```
[root@k8s-master ~]# while sleep 1; do curl rewrite.deanit.cn/; done
# 输出信息
rewrite-myapp-v1-d4659d574-cqc2h
rewrite-myapp-v1-d4659d574-cqc2h
rewrite-myapp-v1-d4659d574-cqc2h
rewrite-myapp-v1-d4659d574-cqc2h
rewrite-myapp-v1-d4659d574-cqc2h
rewrite-myapp-v1-d4659d574-cqc2h
rewrite-myapp-v1-d4659d574-cqc2h
```

08 通过查看Kiali图形化，可以看到请求的整个动态链路图，将Kiali的Service改为NodePort，然后访问任意节点IP+NodePort端口即可，如图15-9所示。

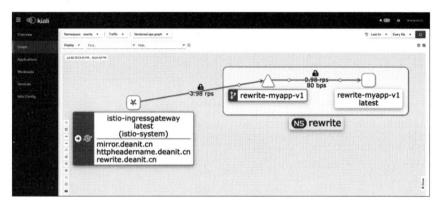

图 15-9　服务请求链路图

15.7.4　重定向

HTTP重定向（Redirect）指的是将请求到原目标服务的流量重定向到给另一个目标服务，客户端请求时不用更改任何方式，即可访问重定向后的目标服务。例如，当前使用的服务发生的变更，需要到新的服务才可以提供访问，在不变更用户原始请求的情况下，通过重定向就可以很好地解决这个问题。

01 创建命名空间：

```
[root@k8s-master ~]# kubectl create ns redirect
namespace/redirect created
```

02 设置Sidecar自动注入：

```
[root@k8s-master ~]# kubectl label namespace redirect istio-injection=enabled
namespace/redirect labeled
```

03 创建资源清单：

```
[root@k8s-master ~]# vim redirect.yaml
---
```

```yaml
# 模拟客户端请求的Pod
apiVersion: apps/v1
kind: Deployment
metadata:
  labels:
    app: client
  name: client
  namespace: redirect
spec:
  replicas: 1
  selector:
    matchLabels:
      app: client
  template:
    metadata:
      labels:
        app: client
    spec:
      containers:
        - image: docker.m.daocloud.io/nginx:latest
          name: client
---
# 入口网关
apiVersion: networking.istio.io/v1beta1
kind: Gateway
metadata:
  name: redirect-gateway
  namespace: redirect
spec:
  selector:
    istio: ingressgateway
  servers:
    - port:
        number: 80
        name: http
        protocol: HTTP
      hosts:
        - "redirect.deanit.cn"
---
# myapp-v1资源
apiVersion: apps/v1
kind: Deployment
metadata:
  name: redirect-myapp-v1
  namespace: redirect
  labels:
    server: redirect-myapp-v1
spec:
  replicas: 1
  selector:
```

```yaml
      matchLabels:
        server: redirect-myapp-v1
    template:
      metadata:
        name: redirect-myapp-v1
        labels:
          server: redirect-myapp-v1
      spec:
        containers:
        - name: redirect-myapp-v1
          image: registry.cn-hangzhou.aliyuncs.com/deanmr/dean:myappv1
          ports:
          - containerPort: 80
---
# myapp-v1服务
apiVersion: v1
kind: Service
metadata:
  name: redirect-myapp-v1
  namespace: redirect
spec:
  type: ClusterIP
  selector:
    server: redirect-myapp-v1
  ports:
  - name: http
    port: 80
    targetPort: 80
    protocol: TCP
---
# myapp-v2资源
apiVersion: apps/v1
kind: Deployment
metadata:
  name: redirect-myapp-v2
  namespace: redirect
  labels:
    server: redirect-myapp-v2
spec:
  replicas: 1
  selector:
    matchLabels:
      server: redirect-myapp-v2
    template:
      metadata:
        name: redirect-myapp-v2
        labels:
          server: redirect-myapp-v2
      spec:
        containers:
```

```yaml
      - name: redirect-myapp-v2
        image: registry.cn-hangzhou.aliyuncs.com/deanmr/dean:myappv2
        ports:
        - containerPort: 80
---
# myapp-v2服务
apiVersion: v1
kind: Service
metadata:
  name: redirect-myapp-v2
  namespace: redirect
spec:
  type: ClusterIP
  selector:
    server: redirect-myapp-v2
  ports:
  - name: http
    port: 80
    targetPort: 80
    protocol: TCP
---
# VirtualService负责权重
apiVersion: networking.istio.io/v1beta1
kind: VirtualService
metadata:
  name: redirect-myapp-vs
  namespace: redirect
spec:
  hosts:
  - "redirect.deanit.cn"
  gateways:
    - redirect-gateway
  http:
  - match:
    - uri:
        prefix: /
    # 重定向
    redirect:
      uri: /hostname.html
      authority: redirect-myapp-v2
      port: 80
```

04 应用资源：

```
[root@k8s-master ~]# kubectl apply -f redirect.yaml
deployment.apps/client created
gateway.networking.istio.io/redirect-gateway created
deployment.apps/redirect-myapp-v1 created
service/redirect-myapp-v1 created
deployment.apps/redirect-myapp-v2 created
service/redirect-myapp-v2 created
```

```
virtualservice.networking.istio.io/redirect-myapp-vs created
```

05 查看资源状态信息：

```
[root@k8s-master ~]# kubectl get pods,svc,gateway,vs,dr -n redirect
NAME                                        READY   STATUS    RESTARTS   AGE
pod/client-68c4b6764d-6fsv7                 2/2     Running   0          70s
pod/redirect-myapp-v1-7dbc577454-v4rkz      2/2     Running   0          70s
pod/redirect-myapp-v2-5bb768cccb-cr5rb      2/2     Running   0          68s

NAME                         TYPE        CLUSTER-IP       EXTERNAL-IP   PORT(S)   AGE
service/redirect-myapp-v1    ClusterIP   10.107.15.228    <none>        80/TCP    70s
service/redirect-myapp-v2    ClusterIP   10.108.41.108    <none>        80/TCP    68s

NAME                                              AGE
gateway.networking.istio.io/redirect-gateway      70s

NAME                                                       GATEWAYS                HOSTS                        AGE
virtualservice.networking.istio.io/redirect-myapp-vs       ["redirect-gateway"]    ["redirect.deanit.cn"]       66s
```

06 进入客户端Pod内，配置hosts文件，通过访问特定域名来注入流量进行测试，可以看到在访问根路径（/）时，流量被重定向到http://redirect-myapp-v2:80/hostname.html，此时HTTP状态码为301：

```
[root@k8s-master ~]# kubectl exec -it -n redirect client-68c4b6764d-6fsv7 sh
kubectl exec [POD] [COMMAND] is DEPRECATED and will be removed in a future version.
Use kubectl exec [POD] -- [COMMAND] instead.
--- 设置hosts，进行域名解析
# echo "192.168.1.21 redirect.deanit.cn">>/etc/hosts

--- 访问域名，查看状态为301重定向
# curl -I redirect.deanit.cn
HTTP/1.1 301 Moved Permanently
location: http://redirect-myapp-v2:80/hostname.html
date: Tue, 30 Jul 2024 15:05:22 GMT
server: envoy
x-envoy-upstream-service-time: 0
transfer-encoding: chunked

--- 循环访问域名，可以看出请求已经重定向到v2服务上
# while sleep 1; do curl -L redirect.deanit.cn; done
--- 输出信息
redirect-myapp-v2-5bb768cccb-cr5rb
redirect-myapp-v2-5bb768cccb-cr5rb
redirect-myapp-v2-5bb768cccb-cr5rb
redirect-myapp-v2-5bb768cccb-cr5rb
redirect-myapp-v2-5bb768cccb-cr5rb
redirect-myapp-v2-5bb768cccb-cr5rb
```

07 通过查看Kiali图形化，可以看到请求的整个动态链路图，将Kiali的Service改为NodePort，然后访问任意节点IP+NodePort端口即可，如图15-10所示。

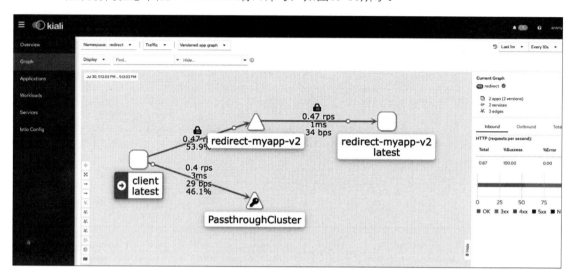

图 15-10　服务请求链路图

15.7.5　流量权重－蓝绿与金丝雀发布

蓝绿部署（Blue-Green Deployment），简单来讲就是在生产环境中部署两套同样的应用，并通过路由进行切换。例如，绿色是线上环境，当我们要发布新版本时，可以在蓝色环境中进行代码更新、测试等操作，确保没有问题后，修改路由规则（如反向代理等）把流量切换到绿色环境。

蓝绿部署是一种热部署方式，目的是尽可能地减少系统下线的时间。蓝绿部署的优点是：可以让用户放心地部署非在线环境，而不用担心部署出错影响生产环境。同时它也提供了快速回滚的能力，比如当我们发现蓝色环境（新版本）出现问题时，可以把流量再切换回绿色环境（旧版本）。蓝绿部署无须停机更新，风险较小。当然，它也有一些不足之处，比如需要两套环境，成本较高，当有未完成的业务（如数据库事务）时，切换版本可能会出现问题。蓝绿部署适合增量更新，在微服务架构中很常用。

对于金丝雀/灰度和蓝绿发布，如果设置某个服务权重为100就是蓝绿发布，如果设置其他的权重（1～99），就是金丝雀发布，并不像Kubernetes的ingress-nginx，还需要开启金丝雀的注解。

01 创建命名空间：

```
[root@k8s-master ~]# kubectl create ns release
namespace/release created
```

02 设置Sidecar自动注入：

```
[root@k8s-master ~]# kubectl label namespace release istio-injection=enabled
```

```
namespace/release labeled
```

03 创建资源清单：

```yaml
[root@k8s-master ~]# vim release.yaml
---
# 入口网关
apiVersion: networking.istio.io/v1beta1
kind: Gateway
metadata:
  name: release-gateway
  namespace: release
spec:
  selector:
    istio: ingressgateway
  servers:
    - port:
        number: 80
        name: http
        protocol: HTTP
      hosts:
        - "release.deanit.cn"
---
# 假设v1服务为旧版本
apiVersion: apps/v1
kind: Deployment
metadata:
  name: myapp-v1
  namespace: release
  labels:
    server: myapp-v1
    app: web
spec:
  replicas: 1
  selector:
    matchLabels:
      server: myapp-v1
      app: web
  template:
    metadata:
      name: myapp-v1
      labels:
        server: myapp-v1
        app: web
    spec:
      containers:
      - name: myapp-v1
        image: registry.cn-hangzhou.aliyuncs.com/deanmr/dean:myappv1
        ports:
        - containerPort: 80
---
```

```yaml
# 假设v2服务为新版本
apiVersion: apps/v1
kind: Deployment
metadata:
  name: myapp-v2
  namespace: release
  labels:
    server: myapp-v2
    app: web
spec:
  replicas: 1
  selector:
    matchLabels:
      server: myapp-v2
      app: web
  template:
    metadata:
      name: myapp-v2
      labels:
        server: myapp-v2
        app: web
    spec:
      containers:
      - name: myapp-v2
        image: registry.cn-hangzhou.aliyuncs.com/deanmr/dean:myappv2
        ports:
        - containerPort: 80
---
# 服务
apiVersion: v1
kind: Service
metadata:
  name: myapp-all
  namespace: release
spec:
  type: ClusterIP
  selector:
    app: web         # <<<----通过app标签匹配v1-v2容器
  ports:
  - name: http
    port: 80
    targetPort: 80
    protocol: TCP
---
# VirtualService负责权重
# DestinationRule根据标签将流量分成不同的子集，已提供VirtualService进行调度，并且设置相关的负载百分比实现精准的控制
apiVersion: networking.istio.io/v1beta1
kind: VirtualService
metadata:
```

```yaml
  name: myapp-vs
  namespace: release
spec:
  hosts:
  - "release.deanit.cn"
  gateways:
    - release-gateway
  http:
  - route:
    - destination:
        host: myapp-all
        subset: v1
      weight: 90          # 权重为90,如果设置为100,那么下面的权重将被删除
    - destination:
        host: myapp-all
        subset: v2
      weight: 10          # 权重为10
---
# DestinationRule负责细分
apiVersion: networking.istio.io/v1beta1
kind: DestinationRule
metadata:
  name: myapp-dr
  namespace: release
spec:
  host: myapp-all
  trafficPolicy:
    loadBalancer:
      simple: RANDOM
  subsets:
  - name: v1
    labels:
      server: myapp-v1      # <<----通过标签容器匹配
  - name: v2
    labels:
      server: myapp-v2      # <<----通过标签容器匹配
```

04 应用资源:

```
[root@k8s-master ~]# kubectl apply -f release.yaml
gateway.networking.istio.io/release-gateway created
deployment.apps/myapp-v1 created
deployment.apps/myapp-v2 created
service/myapp-all created
virtualservice.networking.istio.io/myapp-vs created
destinationrule.networking.istio.io/myapp-dr created
```

05 查看资源状态信息:

```
[root@k8s-master ~]# kubectl get pods,svc,gateway,vs,dr -n release
NAME                        READY   STATUS    RESTARTS   AGE
```

```
pod/myapp-v1-5654d6f9db-t6wnm      2/2     Running     0        45s
pod/myapp-v2-8bf95db78-pm998       2/2     Running     0        45s

NAME                    TYPE         CLUSTER-IP      EXTERNAL-IP   PORT(S)   AGE
service/myapp-all       ClusterIP    10.99.121.12    <none>        80/TCP    45s

NAME                                                  AGE
gateway.networking.istio.io/release-gateway           45s

NAME                                             GATEWAYS              HOSTS                    AGE
virtualservice.networking.istio.io/myapp-vs      ["release-gateway"]   ["release.deanit.cn"]    46s

NAME                                             HOST         AGE
destinationrule.networking.istio.io/myapp-dr     myapp-all    45s
```

06 在k8s-master节点配置hosts进行域名解析,将域名解析到istio-ingressgateway的LoadBalancer分配的IP地址上:

```
[root@k8s-master ~]# echo "192.168.1.21 release.deanit.cn">>/etc/hosts
```

07 进行请求测试,通过访问特定域名来注入流量进行测试,可以看到v1版本的比例比v2要大,因为我们设置的权重为90与10:

```
[root@k8s-master ~]# while sleep 1; do curl release.deanit.cn; done
# 输出信息
Hello MyApp | Version: v1 | <a href="hostname.html">Pod Name</a>
Hello MyApp | Version: v1 | <a href="hostname.html">Pod Name</a>
Hello MyApp | Version: v1 | <a href="hostname.html">Pod Name</a>
Hello MyApp | Version: v1 | <a href="hostname.html">Pod Name</a>
Hello MyApp | Version: v2 | <a href="hostname.html">Pod Name</a>
Hello MyApp | Version: v1 | <a href="hostname.html">Pod Name</a>
Hello MyApp | Version: v1 | <a href="hostname.html">Pod Name</a>
Hello MyApp | Version: v1 | <a href="hostname.html">Pod Name</a>
Hello MyApp | Version: v1 | <a href="hostname.html">Pod Name</a>
Hello MyApp | Version: v1 | <a href="hostname.html">Pod Name</a>
Hello MyApp | Version: v1 | <a href="hostname.html">Pod Name</a>
Hello MyApp | Version: v1 | <a href="hostname.html">Pod Name</a>
Hello MyApp | Version: v1 | <a href="hostname.html">Pod Name</a>
Hello MyApp | Version: v1 | <a href="hostname.html">Pod Name</a>
Hello MyApp | Version: v2 | <a href="hostname.html">Pod Name</a>
```

08 通过查看Kiali图形化,可以看到请求的整个动态请求链路图,将Kiali的Service改为NodePort,然后访问任意节点IP+NodePort端口即可,如图15-11所示。

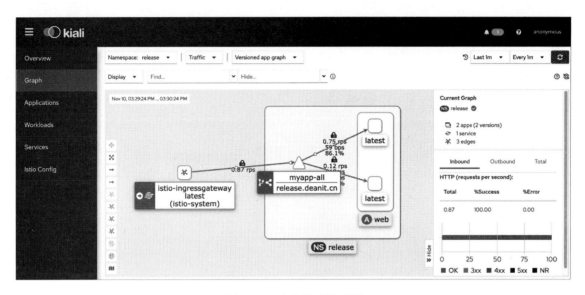

图 15-11　服务请求链路图

15.7.6　超时

超时（TimeOut）是在一个微服务应用中，服务A需要调用服务B。当网络故障造成B出现延迟而无法及时响应时，如果A持续等待，那么它的上游服务也会受到影响，进而使得故障的面积扩散，造成更严重的问题。超时就是一种控制故障范围的机制，相当于一个简单的熔断策略。

网络有时会出现抖动，这将导致通信失败，而重试的主要目的就是解决这一问题。在调用出现失败后进行重试，可以提高服务间交互的成功率和可用性。超时和重试都是微服务应用需要支持的功能，是提高系统弹性的重要保障。

在Istio中，默认超时时间为15s，超时使用timeout标记来设置。当调用上游服务时，如果上游服务一直没有响应，那么就可以设置一个最大等待时间，如果超过这个最大时间，就直接返回，不再继续等待上游服务（快速失败）。其目的是控制故障的范围，避免故障扩散（如果一直等待故障服务的返回，就会把自己的服务拖垮，这样就可以把故障控制在一定范围之内）。服务请求的流程如图15-12所示。

下面我们通过一个实验来演示上述流程。

实验介绍：

（1）Nginx服务设置了超时时间为3s，如果超出这个时间就不再等待，返回超时错误。

（2）Httpd服务设置了响应时间延迟5s，任何请求都需要等待5s后才能返回。

（3）客户端通过访问Nginx服务来反向代理Httpd服务，由于Httpd服务需要5s后才能返回，但Nginx服务只等待3s，因此客户端会提示超时错误。

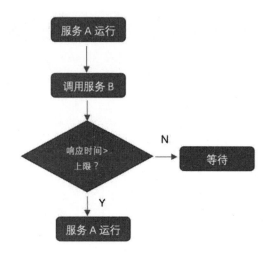

图 15-12　服务请求流程图

（4）客户端→Nginx（设置3s超时，若3s内无响应，则请求失败）→Httpd（故意设置5s延迟，导致Nginx无法及时访问）。

实验步骤如下：

01 创建命名空间：

```
[root@k8s-master ~]# kubectl create ns delay-timeout
namespace/delay-timeout created
```

02 设置Sidecar自动注入：

```
[root@k8s-master ~]# kubectl label ns delay-timeout istio-injection=enabled
namespace/delay-timeout labeled
```

03 创建资源清单：

```
[root@k8s-master ~]# vim delay-timeout.yaml
---
# Nginx网关，代理后端的服务
apiVersion: apps/v1
kind: Deployment
metadata:
  labels:
    app: nginx-gateway
  name: nginx-gateway
  namespace: delay-timeout
spec:
  replicas: 1
  selector:
    matchLabels:
```

```yaml
      app: nginx-gateway
  template:
    metadata:
      labels:
        app: nginx-gateway
    spec:
      containers:
        - image: nginx:latest
          name: nginx-gateway
---
# 后端服务
apiVersion: apps/v1
kind: Deployment
metadata:
  labels:
    app: httpd-backend
  name: httpd-backend
  namespace: delay-timeout
spec:
  replicas: 1
  selector:
    matchLabels:
      app: httpd-backend
  template:
    metadata:
      labels:
        app: httpd-backend
    spec:
      containers:
        - image: httpd:latest
          name: httpd-backend
---
# Nginx网关的svc入口
apiVersion: v1
kind: Service
metadata:
  name: nginx-service
  namespace: delay-timeout
spec:
  selector:
    app: nginx-gateway
  type: ClusterIP
  ports:
    - name: http
      port: 80
      targetPort: 80
      protocol: TCP
---
# 后端服务的svc入口
apiVersion: v1
```

```yaml
kind: Service
metadata:
  name: httpd-service
  namespace: delay-timeout
spec:
  selector:
    app: httpd-backend
  type: ClusterIP
  ports:
    - name: http
      port: 80
      targetPort: 80
      protocol: TCP
---
# 虚拟服务
apiVersion: networking.istio.io/v1beta1
kind: VirtualService
metadata:
  name: nginx-vs
  namespace: delay-timeout
spec:
  hosts:
    - nginx-service
  http:
    - route:
        - destination:
            host: nginx-service
      # 超时3s
      timeout: 3s
---
# 虚拟服务
apiVersion: networking.istio.io/v1beta1
kind: VirtualService
metadata:
  name: httpd-vs
  namespace: delay-timeout
spec:
  hosts:
    - httpd-service
  http:
    - fault:
        # 延迟
        delay:
          # 百分比
          percentage:
            # 值
            value: 100
          # 固定延时5秒
          fixedDelay: 5s
      route:
```

```yaml
    - destination:
        host: httpd-service
---
# 模拟客户端请求的Pod
apiVersion: apps/v1
kind: Deployment
metadata:
  labels:
    app: client
  name: client
  namespace: delay-timeout
spec:
  replicas: 1
  selector:
    matchLabels:
      app: client
  template:
    metadata:
      labels:
        app: client
    spec:
      containers:
        - image: busybox:latest
          name: client
          command: [ "/bin/sh", "-c", "sleep 3600"]
```

04 应用资源:

```
[root@k8s-master ~]# kubectl apply -f delay-timeout.yaml
deployment.apps/nginx-gateway created
deployment.apps/httpd-backend created
service/nginx-service created
service/httpd-service created
virtualservice.networking.istio.io/nginx-vs created
virtualservice.networking.istio.io/httpd-vs created
deployment.apps/client created
```

05 查看资源状态信息:

```
[root@k8s-master ~]# kubectl get pods,svc,gateway,vs,dr -n delay-timeout
NAME                                      READY   STATUS    RESTARTS   AGE
pod/client-74b58898bb-7g8t6               2/2     Running   0          44s
pod/httpd-backend-64f47b6795-4rgc5        2/2     Running   0          46s
pod/nginx-gateway-5978f9fb75-cml9z        2/2     Running   0          46s

NAME                    TYPE        CLUSTER-IP       EXTERNAL-IP   PORT(S)   AGE
service/httpd-service   ClusterIP   10.110.143.122   <none>        80/TCP    46s
service/nginx-service   ClusterIP   10.99.85.57      <none>        80/TCP    46s

NAME                                                   GATEWAYS    HOSTS       AGE
```

```
virtualservice.networking.istio.io/httpd-vs         ["httpd-service"]   45s
virtualservice.networking.istio.io/nginx-vs         ["nginx-service"]   45s
```

06 进入Pod内，配置Nginx反向代理，模拟下游调用：

```
[root@k8s-master ~]# kubectl exec -it -n delay-timeout nginx-gateway-5978f9fb75-cml9z -- sh
--- 创建配置文件，反向代理
# tee /etc/nginx/conf.d/default.conf <<-'EOF'
server {
    listen 80;
    server_name localhost;

    location / {
        proxy_pass http://httpd-service;
        proxy_http_version 1.1;
    }
}
EOF

--- 检测配置文件并重载配置
# nginx -t ; nginx -s reload
nginx: the configuration file /etc/nginx/nginx.conf syntax is ok
nginx: configuration file /etc/nginx/nginx.conf test is successful
2024/07/31 04:57:53 [notice] 46#46: signal process started
# exit
```

07 进入客户端Pod内测试结果，每隔3s，若Nginx服务的超时时间到了而httpd没有响应，则提示返回网关超时错误：

```
[root@k8s-master ~]# kubectl exec -it -n delay-timeout client-74b58898bb-7g8t6 -- sh
/ # while true; do wget -q -O - http://nginx-service; done
wget: server returned error: HTTP/1.1 504 Gateway Timeout
wget: server returned error: HTTP/1.1 504 Gateway Timeout
wget: server returned error: HTTP/1.1 504 Gateway Timeout
wget: server returned error: HTTP/1.1 504 Gateway Timeout
wget: server returned error: HTTP/1.1 504 Gateway Timeout
wget: server returned error: HTTP/1.1 504 Gateway Timeout
wget: server returned error: HTTP/1.1 504 Gateway Timeout
wget: server returned error: HTTP/1.1 504 Gateway Timeout
```

08 通过查看Kiali图形化，可以看到请求的整个动态请求链路图，将Kiali的Service改为NodePort，然后访问任意节点IP+NodePort端口即可，如图15-13所示。

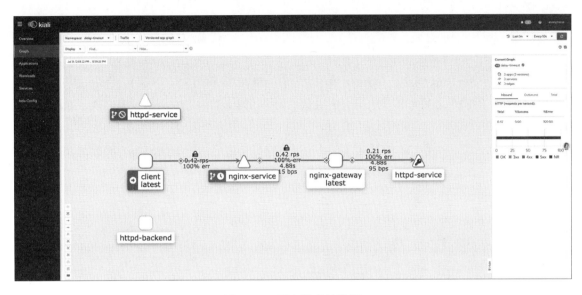

图 15-13　服务请求链路图

15.7.7　重试

重试使用关键字 retires 来设置，包含以下字段：attempts 用来定义尝试的次数，perTryTimeout 用来设置每次尝试等待的时长，retryOn 用来设置遇到什么状态码进行重试。

重试是不断地尝试重新调用请求失败的服务，它通常解决的是由于网络问题导致通信失败的问题，通过重试可以提高系统的稳定性。

重试服务的调用流程如图 15-14 所示。

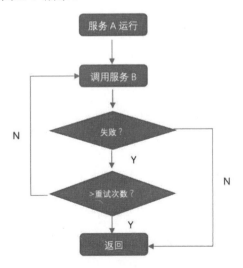

图 15-14　服务调用流程图

下面我们通过一个实验来演示重试服务。

实验介绍：

Nginx服务访问Httpd服务，但Httpd服务由于自身故障错误响应Nginx服务，Nginx服务为了提高容错率，在等待15s之后重新发起第一次重试，如果还是未有响应，则再试第二次，如果还是没有响应，则请求失败。

在大多数场景下，由于故障不是恒定的，而是瞬时出现而后自动恢复的，因此可以通过重试提供服务的可用性。

重试常用参数解释：

- retries参数：可以定义请求失败时的策略，重试策略包括重试次数、超时、重试条件。
- attempts：必选字段，定义重试的次数。
- perTryTimeout：单次重试超时的时间，单位可以是ms、s、m和h。
- retryOn：重试的条件，可以是多个条件，以逗号分隔，例如retryOn: 5xx重试的状态码，只有5xx才会重试。
    - HTTP请求重试条件（route.RertyPlolicy）。
        - attempts: 3：指定最大重试次数。
        - perTryTimeout: 2s：定义每次重试尝试的超时时间。如果单次请求超过这个时间且符合重试条件，则会中止并进行下一次重试，支持时长格式，比如1s表示1秒。
        - retryRemoteLocalities: true：指明是否重试时考虑到远程地域。默认为false，通常用于跨地域时的流量管理。
- retryOn：触发重试的条件。可以指定多种条件，使用逗号分隔：
    - 重试条件1（同x-envoy-retry-on标头）。
        - 5xx：上游主机返回5xx响应码，或者根本未予响应（断开/重试/读取超时）。
        - gateway-error：网关错误，类似于5xx策略，但仅为502、503、504的应用进行重试。
        - connection-failure：在TCP级别与上游服务建立连接超时失败时进行重试。
        - retriable-4xx：上游服务器返回可重复的4xx响应码时进行重试。
        - refused-stream：上游服务使用REFUSED—stream错误码重置时进行重试。
        - retriable-status-codes：上游服务器的响应码与重试策略或者x-envoy-retriable-status-codes标头值中定义的响应码匹配时进行重试。
        - reset：上游主机完全不响应时（disconnect/reset/read超时），Envoy将进行重试。
        - retriable-headers：如果上游服务器响应报文匹配重试策略或x-envoy-retriable-header-names标头中包含的任何标头，则Envoy将尝试重试。
        - envoy-ratelimited：标头中存在x-envoy-ratelimited时进行重试。
    - 重试条件2（同x-envoy-retry-grpc-on标头）。
        - cancelled：gRPC应答标头中的状态码是cancelled时进行重试。

- **deadline-exceeded**：gRPC应答标头中的状态码是deadline-exceeded时进行重试。
- **internal**：gRPC应答标头中的状态码是internal时进行重试。
- **resource-exhausted**：gRPC应答标头中的状态码是resource-exhausted时进行重试。
- **unavailable**：gRPC应答标头中的状态码是unavailable时进行重试。

默认情况下，Envoy不会执行任何类型的重试操作，除非有明确的配置定义。

具体的实验步骤如下。

01 创建命名空间：

```
[root@k8s-master ~]# kubectl create ns retries
namespace/retries created
```

02 设置Sidecar自动注入：

```
[root@k8s-master ~]# kubectl label ns retries istio-injection=enabled
namespace/retries labeled
```

03 创建资源清单：

```
[root@k8s-master ~]# vim retries.yaml
---
# 客户端
apiVersion: apps/v1
kind: Deployment
metadata:
  labels:
    app: client
  name: client
  namespace: retries
spec:
  replicas: 1
  selector:
    matchLabels:
      app: client
  strategy:
    rollingUpdate:
      maxSurge: 25%
      maxUnavailable: 25%
    type: RollingUpdate
  template:
    metadata:
      labels:
        app: client
    spec:
      containers:
      - image: docker.m.daocloud.io/busybox:latest
        name: client-deployment
        command: [ "/bin/sh", "-c", "sleep 3600" ]
```

```yaml
---
# Nginx网关代理
apiVersion: apps/v1
kind: Deployment
metadata:
  labels:
    app: nginx-gateway
  name: nginx-gateway
  namespace: retries
spec:
  replicas: 1
  selector:
    matchLabels:
      app: nginx-gateway
  strategy:
    rollingUpdate:
      maxSurge: 25%
      maxUnavailable: 25%
    type: RollingUpdate
  template:
    metadata:
      labels:
        app: nginx-gateway
    spec:
      containers:
        - image: docker.m.daocloud.io/nginx:latest
          name: nginx-gateway
---
# 后端服务
apiVersion: apps/v1
kind: Deployment
metadata:
  labels:
    app: httpd-backend
  name: httpd-backend
  namespace: retries
spec:
  replicas: 1
  selector:
    matchLabels:
      app: httpd-backend
  strategy:
    rollingUpdate:
      maxSurge: 25%
      maxUnavailable: 25%
    type: RollingUpdate
  template:
    metadata:
      labels:
        app: httpd-backend
```

```yaml
  spec:
    containers:
      - image: httpd:latest
        name: httpd-backend
---
# Nginx网关的svc入口
apiVersion: v1
kind: Service
metadata:
  name: nginx-service
  namespace: retries
spec:
  selector:
    app: nginx-gateway
  type: ClusterIP
  ports:
  - name: http
    port: 80
    targetPort: 80
    protocol: TCP
---
# 后端服务的svc入口
apiVersion: v1
kind: Service
metadata:
  name: httpd-service
  namespace: retries
spec:
  selector:
    app: httpd-backend
  type: ClusterIP
  ports:
  - name: http
    port: 80
    targetPort: 80
    protocol: TCP
---
# Nginx网关的虚拟服务
apiVersion: networking.istio.io/v1beta1
kind: VirtualService
metadata:
  name: nginx-vs
  namespace: retries
spec:
  hosts:
  - nginx-service
  http:
  - route:
    - destination:
        host: nginx-service
```

```yaml
    retries:
      attempts: 3
      perTryTimeout: 5s
---
# 后端服务的虚拟服务
apiVersion: networking.istio.io/v1beta1
kind: VirtualService
metadata:
  name: httpd-vs
  namespace: retries
spec:
  hosts:
  - httpd-service
  http:
  - fault:
      abort:
        percentage:
          value: 100
        # 返回请求状态码
        httpStatus: 503
    route:
    - destination:
        host: httpd-service
```

04 应用资源：

```
[root@k8s-master ~]# kubectl apply -f retries.yaml
deployment.apps/client created
deployment.apps/nginx-gateway created
deployment.apps/httpd-backend created
service/nginx-service created
service/httpd-service created
virtualservice.networking.istio.io/nginx-vs created
virtualservice.networking.istio.io/httpd-vs created
```

05 查看资源状态信息：

```
[root@k8s-master ~]# kubectl get pods,svc,gateway,vs,dr -n retries
NAME                                        READY   STATUS    RESTARTS   AGE
pod/client-b4c94dc67-2mwsh                  2/2     Running   0          37s
pod/httpd-backend-64f47b6795-ndbh4          2/2     Running   0          35s
pod/nginx-gateway-678cd9fd66-tm682          2/2     Running   0          36s

NAME                    TYPE        CLUSTER-IP     EXTERNAL-IP   PORT(S)   AGE
service/httpd-service   ClusterIP   10.96.66.90    <none>        80/TCP    33s
service/nginx-service   ClusterIP   10.105.27.70   <none>        80/TCP    34s

NAME                                            GATEWAYS   HOSTS                AGE
virtualservice.networking.istio.io/httpd-vs                ["httpd-service"]    32s
virtualservice.networking.istio.io/nginx-vs                ["nginx-service"]    33s
```

06 进入Pod内配置Nginx反向代理Httpd以完成上下游调用的效果：

```
[root@k8s-master ~]# kubectl exec -it -n retries nginx-gateway-678cd9fd66-tm682 -- sh
--- 创建配置文件，反向代理
# tee /etc/nginx/conf.d/default.conf <<-'EOF'
server {
    listen 80;
    server_name localhost;

    location / {
        proxy_pass http://httpd-service;
        proxy_http_version 1.1;
    }
}
EOF

--- 检测配置文件并重载配置
# nginx -t ; nginx -s reload
nginx: the configuration file /etc/nginx/nginx.conf syntax is ok
nginx: configuration file /etc/nginx/nginx.conf test is successful
2024/07/31 05:34:16 [notice] 39#39: signal process started
# exit
```

07 进入客户端Pod内进行请求访问测试：

```
[root@k8s-master ~]# kubectl exec -it -n retries client-b4c94dc67-2mwsh -- sh
/ # wget -q -O - http://nginx-service
wget: server returned error: HTTP/1.1 503 Service Unavailable
```

08 查看边车日志，自己请求一次，3次为重试请求，一共请求了4次，流出outbound，流入inbound，分别为4次，所以通过如下日志可以看出一共有8条请求日志：

```
[root@k8s-master ~]# kubectl logs -f nginx-gateway-678cd9fd66-tm682 -c istio-proxy -n retries
# 输出信息
2024-07-31T05:33:08.852396Z     info    FLAG: --concurrency="2"
2024-07-31T05:33:08.852425Z     info    FLAG: --domain="retries.svc.cluster.local"
2024-07-31T05:33:08.852429Z     info    FLAG: --help="false"
2024-07-31T05:33:08.852431Z     info    FLAG: --log_as_json="false"
2024-07-31T05:33:08.852432Z     info    FLAG: --log_caller=""
2024-07-31T05:33:08.852434Z     info    FLAG: --log_output_level="default:info"
2024-07-31T05:33:08.852435Z     info    FLAG: --log_rotate=""
2024-07-31T05:33:08.852437Z     info    FLAG: --log_rotate_max_age="30"
2024-07-31T05:33:08.852439Z     info    FLAG: --log_rotate_max_backups="1000"
2024-07-31T05:33:08.852441Z     info    FLAG: --log_rotate_max_size="104857600"
2024-07-31T05:33:08.852443Z     info    FLAG: --log_stacktrace_level="default:none"
2024-07-31T05:33:08.852450Z     info    FLAG: --log_target="[stdout]"
2024-07-31T05:33:08.852454Z     info    FLAG: --meshConfig="./etc/istio/config/mesh"
```

```
2024-07-31T05:33:08.852466Z     info    FLAG: --outlierLogPath=""
2024-07-31T05:33:08.852468Z     info    FLAG: --proxyComponentLogLevel="misc:error"
2024-07-31T05:33:08.852469Z     info    FLAG: --proxyLogLevel="warning"
2024-07-31T05:33:08.852471Z     info    FLAG: --serviceCluster="istio-proxy"
2024-07-31T05:33:08.852472Z     info    FLAG: --stsPort="0"
2024-07-31T05:33:08.852474Z     info    FLAG: --templateFile=""
2024-07-31T05:33:08.852476Z     info    FLAG:
--tokenManagerPlugin="GoogleTokenExchange"
2024-07-31T05:33:08.852480Z     info    FLAG: --vklog="0"
2024-07-31T05:33:08.852483Z     info    Version
1.17.8-a781f9ee6c511d8f22140d8990c31e577b2a9676-Clean
2024-07-31T05:33:08.854231Z     info    Maximum file descriptors (ulimit -n): 1048576
2024-07-31T05:33:08.854464Z     info    Proxy role     ips=[10.244.169.155]
type=sidecar id=nginx-gateway-678cd9fd66-tm682.retries domain=retries.svc.cluster.local
2024-07-31T05:33:08.854586Z     info    Apply proxy config from env {}

2024-07-31T05:33:08.858505Z     info    Effective config: binaryPath:
/usr/local/bin/envoy
  concurrency: 2
  configPath: ./etc/istio/proxy
  controlPlaneAuthPolicy: MUTUAL_TLS
  discoveryAddress: istiod.istio-system.svc:15012
  drainDuration: 45s
  proxyAdminPort: 15000
  serviceCluster: istio-proxy
  statNameLength: 189
  statusPort: 15020
  terminationDrainDuration: 5s
  tracing:
    zipkin:
      address: zipkin.istio-system:9411

2024-07-31T05:33:08.858537Z     info    JWT policy is third-party-jwt
2024-07-31T05:33:08.858541Z     info    using credential fetcher of JWT type in
cluster.local trust domain
2024-07-31T05:33:09.060670Z     info    Workload SDS socket not found. Starting Istio
SDS Server
2024-07-31T05:33:09.060712Z     info    CA Endpoint istiod.istio-system.svc:15012,
provider Citadel
2024-07-31T05:33:09.060670Z     info    Opening status port 15020
2024-07-31T05:33:09.060738Z     info    Using CA istiod.istio-system.svc:15012 cert
with certs: var/run/secrets/istio/root-cert.pem
2024-07-31T05:33:09.078813Z     info    ads     All caches have been synced up in
228.555572ms, marking server ready
2024-07-31T05:33:09.079137Z     info    xdsproxy       Initializing with upstream
address "istiod.istio-system.svc:15012" and cluster "Kubernetes"
2024-07-31T05:33:09.079190Z     info    sds     Starting SDS grpc server
2024-07-31T05:33:09.079499Z     info    starting Http service at 127.0.0.1:15004
2024-07-31T05:33:09.080657Z     info    Pilot SAN: [istiod.istio-system.svc]
2024-07-31T05:33:09.081761Z     info    Starting proxy agent
```

```
2024-07-31T05:33:09.081823Z     info    starting
2024-07-31T05:33:09.081855Z     info    Envoy command: [-c
etc/istio/proxy/envoy-rev.json --drain-time-s 45 --drain-strategy immediate
--local-address-ip-version v4 --file-flush-interval-msec 1000 --disable-hot-restart
--allow-unknown-static-fields --log-format %Y-%m-%dT%T.%fZ     %l     envoy %n %g:%#  %v
thread=%t -l warning --component-log-level misc:error --concurrency 2]
2024-07-31T05:33:09.163020Z     info    xdsproxy        connected to upstream XDS
server: istiod.istio-system.svc:15012
2024-07-31T05:33:09.208209Z     info    ads     ADS: new connection for
node:nginx-gateway-678cd9fd66-tm682.retries-1
2024-07-31T05:33:09.209341Z     info    ads     ADS: new connection for
node:nginx-gateway-678cd9fd66-tm682.retries-2
2024-07-31T05:33:09.483983Z     info    cache   generated new workload certificate
latency=404.769315ms ttl=23h59m59.516032304s
2024-07-31T05:33:09.484019Z     info    cache   Root cert has changed, start rotating
root cert
2024-07-31T05:33:09.484036Z     info    ads     XDS: Incremental Pushing:0
ConnectedEndpoints:2 Version:
2024-07-31T05:33:09.484093Z     info    cache   returned workload trust anchor from
cache    ttl=23h59m59.515909129s
2024-07-31T05:33:09.484110Z     info    cache   returned workload certificate from
cache    ttl=23h59m59.515891639s
2024-07-31T05:33:09.484228Z     info    cache   returned workload trust anchor from
cache    ttl=23h59m59.515777668s
2024-07-31T05:33:09.484473Z     info    ads     SDS: PUSH request for
node:nginx-gateway-678cd9fd66-tm682.retries resources:1 size:4.0kB resource:default
2024-07-31T05:33:09.484717Z     info    ads     SDS: PUSH request for
node:nginx-gateway-678cd9fd66-tm682.retries resources:1 size:1.1kB resource:ROOTCA
2024-07-31T05:33:09.484828Z     info    cache   returned workload trust anchor from
cache    ttl=23h59m59.515175051s
2024-07-31T05:33:09.619449Z     info    Readiness succeeded in 774.506149ms
2024-07-31T05:33:09.619824Z     info    Envoy proxy is ready

    [2024-07-31T05:35:43.865Z] "GET / HTTP/1.1" 503 FI fault_filter_abort - "-" 0 18 0 -
"-" "Wget" "1d787578-1dcb-911d-ac12-3da780b8ed5c" "httpd-service" "-"
outbound|80||httpd-service.retries.svc.cluster.local - 10.96.66.90:80
10.244.169.155:41084 - -
    [2024-07-31T05:35:43.865Z] "GET / HTTP/1.1" 503 - via_upstream - "-" 0 18 0 0 "-" "Wget"
"1d787578-1dcb-911d-ac12-3da780b8ed5c" "nginx-service" "10.244.169.155:80" inbound|80||
127.0.0.6:57631 10.244.169.155:80 10.244.169.168:60112
outbound_.80_._.nginx-service.retries.svc.cluster.local default
    [2024-07-31T05:35:43.873Z] "GET / HTTP/1.1" 503 FI fault_filter_abort - "-" 0 18 0 -
"-" "Wget" "1d787578-1dcb-911d-ac12-3da780b8ed5c" "httpd-service" "-"
outbound|80||httpd-service.retries.svc.cluster.local - 10.96.66.90:80
10.244.169.155:41096 - -
    [2024-07-31T05:35:43.873Z] "GET / HTTP/1.1" 503 - via_upstream - "-" 0 18 0 0 "-" "Wget"
"1d787578-1dcb-911d-ac12-3da780b8ed5c" "nginx-service" "10.244.169.155:80" inbound|80||
127.0.0.6:57631 10.244.169.155:80 10.244.169.168:60122
outbound_.80_._.nginx-service.retries.svc.cluster.local default
    [2024-07-31T05:35:43.923Z] "GET / HTTP/1.1" 503 FI fault_filter_abort - "-" 0 18 0 -
```

```
"-" "Wget" "1d787578-1dcb-911d-ac12-3da780b8ed5c" "httpd-service" "-"
outbound|80||httpd-service.retries.svc.cluster.local - 10.96.66.90:80
10.244.169.155:41102 - -
        [2024-07-31T05:35:43.922Z] "GET / HTTP/1.1" 503 - via_upstream - "-" 0 18 0 0 "-" "Wget"
"1d787578-1dcb-911d-ac12-3da780b8ed5c" "nginx-service" "10.244.169.155:80" inbound|80||
127.0.0.6:57631 10.244.169.155:80 10.244.169.168:60124
outbound_.80_._.nginx-service.retries.svc.cluster.local default
        [2024-07-31T05:35:44.025Z] "GET / HTTP/1.1" 503 FI fault_filter_abort - "-" 0 18 0 -
"-" "Wget" "1d787578-1dcb-911d-ac12-3da780b8ed5c" "httpd-service" "-"
outbound|80||httpd-service.retries.svc.cluster.local - 10.96.66.90:80
10.244.169.155:41112 - -
        [2024-07-31T05:35:44.024Z] "GET / HTTP/1.1" 503 - via_upstream - "-" 0 18 0 0 "-" "Wget"
"1d787578-1dcb-911d-ac12-3da780b8ed5c" "nginx-service" "10.244.169.155:80" inbound|80||
127.0.0.6:57631 10.244.169.155:80 10.244.169.168:60130
outbound_.80_._.nginx-service.retries.svc.cluster.local default
```

09 进入客户端Pod内进行请求访问测试：

```
[root@k8s-master ~]# kubectl exec -it -n retries client-b4c94dc67-2mwsh -- sh
/ # while true; do wget -q -O - http://nginx-service; done
wget: server returned error: HTTP/1.1 503 Service Unavailable
wget: server returned error: HTTP/1.1 503 Service Unavailable
wget: server returned error: HTTP/1.1 503 Service Unavailable
wget: server returned error: HTTP/1.1 503 Service Unavailable
wget: server returned error: HTTP/1.1 503 Service Unavailable
```

10 通过查看Kiali图形化，可以看到请求的整个动态请求链路图，将Kiali的Service改为NodePort，然后访问任意节点IP+NodePort端口即可，如图15-15所示。

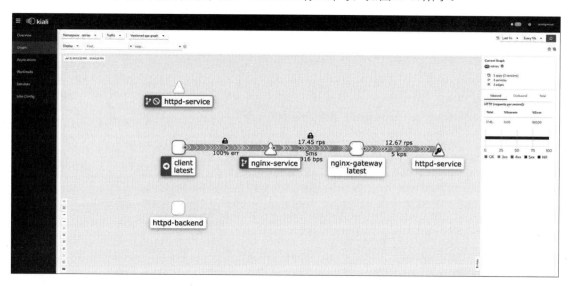

图15-15　服务请求链路图

15.7.8 断路器/熔断

断路器也称为服务熔断，在多个服务调用时，服务A依赖服务B，服务B依赖服务C，如果服务C响应时间过长或者不可用，则会让服务B占用太多系统资源，而服务A也依赖服务B，同时占用大量的系统资源，造成系统雪崩的情况出现。Istio断路器通过网格中的边车对流量进行拦截判断处理，避免了在代码中侵入控制逻辑，可以非常方便地实现服务熔断。

简单来说，电路中的断路器就像带有保险丝的保护装置，一旦电压不稳或短路，系统就会自动跳闸切断电流，以免电流过大烧坏电器。在分布式系统中也有类似的功能，这就是熔断。它的工作原理与电路系统中的跳闸非常类似，当下游服务出现错误时，通过重试发现服务暂时无法恢复，再进行无休止的重试已经没有意义了。此时上游服务为保护自己不受牵连，会切断对下游服务的调用。它是分布式系统应该具有的重要的弹性能力。

图15-16展示了一个熔断器开启和关闭的例子。当下游不可用时，打开熔断，直接返回，不再调用后端服务。

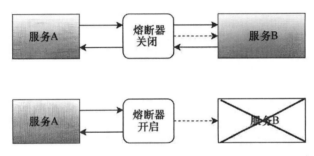

图 15-16　熔断服务请求链路图

接下来将详细介绍如何设置和使用Istio断路器实现熔断保护。

重试常用参数解释：

```
apiVersion: networking.istio.io/v1beta1
kind: DestinationRule
metadata:
  # 规则名称
  name: httpbin
spec:
  # 后端svc
  host: httpbin
  trafficPolicy:
    # 上游主机的连接池配置
    connectionPool:
      tcp:
        # 连接池最大连接数
        maxConnections: 1
        # tcp连接超时时间，默认为秒，可改为ms
        connectTimeout: 30ms
```

```
            # 如果在套接字上设置SO_KEEPALIVE，可以确保TCP存活
            tcpKeepalive:
              # 发送keep-alive探测前连接存在的空闲时间。默认值是使用系统的配置
              time: 7200s
              # 探测活动之间的时间间隔，默认值是使用系统的配置
              interval: 75s
        http:
          # HTTP请求pending状态的最大请求数，从应用容器发来的HTTP请求的最大等待转发数，默认为1024
          http1MaxPendingRequests: 1
          # 在一定时间内限制对后端服务发起的最大请求数，如果超过此配置，就会出现限流，后端请求的最大
数量默认是1024
          maxRequestsPerConnection: 1
          # 在给定时间内，集群中所有主机都可以执行的最大重试次数，默认为3次
          maxRetries: 3
      outlierDetection:
        # 拒绝连接的最大失败次数，如果超过这一配置的数量，服务就会被移出连接池，默认值为5
        consecutiveErrors: 5
        # 触发熔断的时间间隔，在interval时间间隔内，达到consecutiveErrors即触发熔断
        interval: 10s
        # 主机每次被移除后的隔离时间等于被移除的次数和最小移除时间的乘积。这样的实现让系统能够自动增
加不健康上游服务实例的隔离时间。默认值为30s
        # 熔断时长，默认为30s，最小的移除时间长度
        baseEjectionTime: 30s
        # 熔断连接最大百分比，服务在负载均衡池中被拒绝访问的最大百分比，负载均衡池中最多有多大比例被
剔除，默认为10%
        maxEjectionPercent: 10
        # 最小健康百分比阈值
        minHealthPercent: 50
```

设置和使用Istio断路器进行熔断保护的具体操作步骤如下。

01 创建命名空间：

```
[root@k8s-master ~]# kubectl create ns fusing
namespace/fusing created
```

02 设置Sidecar自动注入：

```
[root@k8s-master ~]# kubectl label ns fusing istio-injection=enabled
namespace/fusing labeled
```

03 创建资源清单，如果你的Istio启用了双向TLS身份验证，则必须在应用目标规则之前将TLS流量策略 mode: ISTIO_MUTUAL添加到DestinationRule，否则请求将产生503错误。

```
[root@k8s-master ~]# vim fusing.yaml
apiVersion: v1
kind: ServiceAccount
metadata:
  name: httpbin
  namespace: fusing
---
```

```yaml
apiVersion: v1
kind: Service
metadata:
  name: httpbin
  namespace: fusing
  labels:
    app: httpbin
    service: httpbin
spec:
  ports:
    - name: http
      port: 8000
      targetPort: 80
  selector:
    app: httpbin
---
apiVersion: apps/v1
kind: Deployment
metadata:
  name: httpbin
  namespace: fusing
spec:
  replicas: 1
  selector:
    matchLabels:
      app: httpbin
      version: v1
  template:
    metadata:
      labels:
        app: httpbin
        version: v1
    spec:
      serviceAccountName: httpbin
      containers:
        - image: docker.m.daocloud.io/kennethreitz/httpbin
          imagePullPolicy: IfNotPresent
          name: httpbin
          ports:
            - containerPort: 80
---
apiVersion: networking.istio.io/v1beta1
kind: DestinationRule
metadata:
  name: httpbin
  namespace: fusing
spec:
  host: httpbin
  trafficPolicy:
    connectionPool:
```

```yaml
      http:
        http1MaxPendingRequests: 1
        maxRequestsPerConnection: 1
      tcp:
        maxConnections: 1
    outlierDetection:
      baseEjectionTime: 3m
      # 警告: consecutiveErrors已弃用
      # 使用consecutiveGatewayErrors或consecutive5xxErrors代替
      consecutiveErrors: 1
      interval: 1s
      maxEjectionPercent: 100
---
apiVersion: apps/v1
kind: Deployment
metadata:
  name: fortio-deploy
  namespace: fusing
spec:
  replicas: 1
  selector:
    matchLabels:
      app: fortio
  template:
    metadata:
      annotations:
        # This annotation causes Envoy to serve cluster.outbound statistics via 15000/stats
        # in addition to the stats normally served by Istio. The Circuit Breaking example task
        # gives an example of inspecting Envoy stats via proxy config.
        proxy.istio.io/config: |-
          proxyStatsMatcher:
            inclusionPrefixes:
            - "cluster.outbound"
            - "cluster_manager"
            - "listener_manager"
            - "server"
            - "cluster.xds-grpc"
      labels:
        app: fortio
    spec:
      containers:
      - name: fortio
        image: docker.m.daocloud.io/fortio/fortio:latest_release
        imagePullPolicy: Always
        ports:
        - containerPort: 8080
          name: http-fortio
        - containerPort: 8079
```

```yaml
      name: grpc-ping
---
apiVersion: v1
kind: Service
metadata:
  name: fortio
  namespace: fusing
  labels:
    app: fortio
    service: fortio
spec:
  ports:
  - port: 8080
    name: http
  selector:
    app: fortio
```

04 应用资源：

```
[root@k8s-master ~]# kubectl apply -f fusing.yaml
serviceaccount/httpbin created
service/httpbin created
deployment.apps/httpbin created
destinationrule.networking.istio.io/httpbin created
deployment.apps/fortio-deploy created
service/fortio created
```

05 查看资源状态信息：

```
[root@k8s-master ~]# kubectl get pods,svc,dr,sa -n fusing
NAME                                    READY   STATUS    RESTARTS   AGE
pod/fortio-deploy-745fcf8bb-tfjpb       2/2     Running   0          13m
pod/httpbin-8675df8686-x2ltk            2/2     Running   0          13m

NAME              TYPE        CLUSTER-IP       EXTERNAL-IP   PORT(S)    AGE
service/fortio    ClusterIP   10.98.12.184     <none>        8080/TCP   13m
service/httpbin   ClusterIP   10.107.190.211   <none>        8000/TCP   14m

NAME                                                HOST      AGE
destinationrule.networking.istio.io/httpbin         httpbin   13m

NAME                            SECRETS   AGE
serviceaccount/default          1         23m
serviceaccount/httpbin          1         14m
```

06 进入客户端Pod内并使用Fortio工具调用httpbin服务，-curl参数表明发送一次调用，可以看到调用后端服务的请求已经成功：

```
--- 查询出来fortio的名字
[root@k8s-master ~]# FORTIO_POD=$(kubectl get pod -n fusing | grep fortio | awk '{ print $1 }')
```

```
--- 访问
[root@k8s-master ~]# kubectl exec -it $FORTIO_POD -n fusing -c fortio -- /usr/bin/fortio
load -curl http://httpbin:8000/get
# 输出信息
04:16:09.187 r1 [INF] scli.go:123> Starting, command="Φορτίο", version="1.60.3
h1:adR0uf/69M5xxKaMLAautVf9FIVkEpMwuEWyMaaSnI0= go1.20.10 amd64 linux"
HTTP/1.1 200 OK
server: envoy
date: Sun, 04 Aug 2024 04:16:09 GMT
content-type: application/json
content-length: 592
access-control-allow-origin: *
access-control-allow-credentials: true
x-envoy-upstream-service-time: 22

{
  "args": {},
  "headers": {
    "Host": "httpbin:8000",
    "User-Agent": "fortio.org/fortio-1.60.3",
    "X-B3-Parentspanid": "49ecaa325291ce54",
    "X-B3-Sampled": "1",
    "X-B3-Spanid": "ad16260469dfd4a0",
    "X-B3-Traceid": "e2d27b34a7eb245549ecaa325291ce54",
    "X-Envoy-Attempt-Count": "1",
    "X-Forwarded-Client-Cert":
"By=spiffe://cluster.local/ns/fusing/sa/httpbin;Hash=3d7f1753bb967c8ebecf86687b74c0ae4
909fc64782c92c214f91a66d7d0e833;Subject=\"\";URI=spiffe://cluster.local/ns/fusing/sa/d
efault"
  },
  "origin": "127.0.0.6",
  "url": "http://httpbin:8000/get"
}
```

07 触发熔断器：

```
[root@k8s-master ~]# kubectl exec -it $FORTIO_POD -n fusing -c fortio -- /usr/bin/fortio
load -c 2 -qps 0 -n 20 -loglevel Warning http://httpbin:8000/get
# 输出信息
04:16:28.465 r1 [INF] logger.go:254> Log level is now 3 Warning (was 2 Info)
Fortio 1.60.3 running at 0 queries per second, 4->4 procs, for 20 calls:
http://httpbin:8000/get
Starting at max qps with 2 thread(s) [gomax 4] for exactly 20 calls (10 per thread
+ 0)
04:16:28.476 r12 [WRN] http_client.go:1104> Non ok http code, code=503,
status="HTTP/1.1 503", thread=1, run=0
04:16:28.488 r11 [WRN] http_client.go:1104> Non ok http code, code=503,
status="HTTP/1.1 503", thread=0, run=0
04:16:28.491 r12 [WRN] http_client.go:1104> Non ok http code, code=503,
status="HTTP/1.1 503", thread=1, run=0
04:16:28.508 r11 [WRN] http_client.go:1104> Non ok http code, code=503,
```

```
status="HTTP/1.1 503", thread=0, run=0
    04:16:28.513 r12 [WRN] http_client.go:1104> Non ok http code, code=503,
status="HTTP/1.1 503", thread=1, run=0
    04:16:28.517 r11 [WRN] http_client.go:1104> Non ok http code, code=503,
status="HTTP/1.1 503", thread=0, run=0
    04:16:28.522 r12 [WRN] http_client.go:1104> Non ok http code, code=503,
status="HTTP/1.1 503", thread=1, run=0
    Ended after 71.242138ms : 20 calls. qps=280.73
    Aggregated Function Time : count 20 avg 0.007010968 +/- 0.005702 min 0.000670417 max
0.020558719 sum 0.140219361
    # range, mid point, percentile, count
    >= 0.000670417 <= 0.001 , 0.000835208 , 15.00, 3
    > 0.001 <= 0.002 , 0.0015 , 25.00, 2
    > 0.003 <= 0.004 , 0.0035 , 30.00, 1
    > 0.004 <= 0.005 , 0.0045 , 50.00, 4
    > 0.006 <= 0.007 , 0.0065 , 60.00, 2
    > 0.007 <= 0.008 , 0.0075 , 70.00, 2
    > 0.008 <= 0.009 , 0.0085 , 80.00, 2
    > 0.014 <= 0.016 , 0.015 , 90.00, 2
    > 0.018 <= 0.02 , 0.019 , 95.00, 1
    > 0.02 <= 0.0205587 , 0.0202794 , 100.00, 1
    # target 50% 0.005
    # target 75% 0.0085
    # target 90% 0.016
    # target 99% 0.020447
    # target 99.9% 0.0205475
    Error cases : count 7 avg 0.0029236547 +/- 0.002892 min 0.000670417 max 0.00842827
sum 0.020465583
    # range, mid point, percentile, count
    >= 0.000670417 <= 0.001 , 0.000835208 , 42.86, 3
    > 0.001 <= 0.002 , 0.0015 , 71.43, 2
    > 0.006 <= 0.007 , 0.0065 , 85.71, 1
    > 0.008 <= 0.00842827 , 0.00821414 , 100.00, 1
    # target 50% 0.00125
    # target 75% 0.00625
    # target 90% 0.00812848
    # target 99% 0.00839829
    # target 99.9% 0.00842527
    # Socket and IP used for each connection:
    [0]   4 socket used, resolved to 10.107.190.211:8000, connection timing : count 4 avg
0.000150294 +/- 3.856e-05 min 0.000101799 max 0.00020931 sum 0.000601176
    [1]   5 socket used, resolved to 10.107.190.211:8000, connection timing : count 5 avg
0.0001762122 +/- 3.37e-05 min 0.000141996 max 0.000222674 sum 0.000881061
    Connection time (s) : count 9 avg 0.000164693 +/- 3.818e-05 min 0.000101799 max
0.000222674 sum 0.001482237
    Sockets used: 9 (for perfect keepalive, would be 2)
    Uniform: false, Jitter: false, Catchup allowed: true
    IP addresses distribution:
    10.107.190.211:8000: 9
    Code 200 : 13 (65.0 %)
```

```
Code 503 : 7 (35.0 %)
Response Header Sizes : count 20 avg 149.7 +/- 109.9 min 0 max 231 sum 2994
Response Body/Total Sizes : count 20 avg 618.85 +/- 277.3 min 241 max 823 sum 12377
All done 20 calls (plus 0 warmup) 7.011 ms avg, 280.7 qps
```

参数解析：

- **kubectl exec**：这是执行命令的开始。
- **-it**：-i或--stdin保持STDIN开放，这允许用户与执行的命令进行交互（比如通过终端输入）。-t或--tty分配一个伪终端，这通常用于确保输出的格式正确，特别是当命令是交互式时。
- **$FORTIO_POD**：这是一个环境变量，它应该被替换为实际要执行命令的Pod名称。在实际使用时，需要确保这个环境变量已经设置，或者可以直接将Pod名称替换为这里的$FORTIO_POD。
- **-n fusing**：指定了Pod所在的命名空间是fusing。如果没有在默认命名空间（通常是default）中运行Pod，那么需要指定正确的命名空间。
- **-c fortio**：指定了Pod中的容器名称是fortio。一个Pod可以包含多个容器，-c选项允许指定要执行命令的容器。
- **--**：这个符号告诉kubectl exec命令选项到此为止，后面的内容应该被解释为传递给Pod中命令的参数。
- **/usr/bin/fortio load**：这是要在Pod的Fortio容器中执行的命令。/usr/bin/fortio是Fortio工具的路径，load是Fortio的一个子命令，用于执行HTTP负载测试。
- **-c 2**：指定并发客户端数为2。
- **-qps 0**：指定每秒查询数（Queries Per Second）为0，这实际上意味着没有固定的QPS速率，而是由-n参数控制总的请求数。
- **-n 20**：指定总共发送20个请求。
- **-loglevel Warning**：设置日志级别为Warning，意味着只有警告及以上级别的日志会被输出。
- **http://httpbin:8000/get**：指定了要测试的HTTP服务的URL。这里假设在同一个集群中有另一个服务httpbin，它监听在8000端口上，并且有一个/get端点。

08 将并发连接数提高到3个：

```
[root@k8s-master ~]# kubectl exec -it $FORTIO_POD -n fusing -c fortio -- /usr/bin/fortio
load -c 3 -qps 0 -n 30 -loglevel Warning http://httpbin:8000/get
# 输出信息
04:16:46.481 r1 [INF] logger.go:254> Log level is now 3 Warning (was 2 Info)
Fortio 1.60.3 running at 0 queries per second, 4->4 procs, for 30 calls:
http://httpbin:8000/get
Starting at max qps with 3 thread(s) [gomax 4] for exactly 30 calls (10 per thread
+ 0)
04:16:46.486 r34 [WRN] http_client.go:1104> Non ok http code, code=503,
status="HTTP/1.1 503", thread=0, run=0
04:16:46.487 r35 [WRN] http_client.go:1104> Non ok http code, code=503,
```

```
status="HTTP/1.1 503", thread=1, run=0
    04:16:46.489 r35 [WRN] http_client.go:1104> Non ok http code, code=503,
status="HTTP/1.1 503", thread=1, run=0
    04:16:46.492 r35 [WRN] http_client.go:1104> Non ok http code, code=503,
status="HTTP/1.1 503", thread=1, run=0
    04:16:46.495 r36 [WRN] http_client.go:1104> Non ok http code, code=503,
status="HTTP/1.1 503", thread=2, run=0
    04:16:46.505 r35 [WRN] http_client.go:1104> Non ok http code, code=503,
status="HTTP/1.1 503", thread=1, run=0
    04:16:46.507 r35 [WRN] http_client.go:1104> Non ok http code, code=503,
status="HTTP/1.1 503", thread=1, run=0
    04:16:46.509 r35 [WRN] http_client.go:1104> Non ok http code, code=503,
status="HTTP/1.1 503", thread=1, run=0
    04:16:46.511 r35 [WRN] http_client.go:1104> Non ok http code, code=503,
status="HTTP/1.1 503", thread=1, run=0
    04:16:46.515 r35 [WRN] http_client.go:1104> Non ok http code, code=503,
status="HTTP/1.1 503", thread=1, run=0
    04:16:46.519 r35 [WRN] http_client.go:1104> Non ok http code, code=503,
status="HTTP/1.1 503", thread=1, run=0
    04:16:46.521 r34 [WRN] http_client.go:1104> Non ok http code, code=503,
status="HTTP/1.1 503", thread=0, run=0
    04:16:46.522 r34 [WRN] http_client.go:1104> Non ok http code, code=503,
status="HTTP/1.1 503", thread=0, run=0
    04:16:46.525 r34 [WRN] http_client.go:1104> Non ok http code, code=503,
status="HTTP/1.1 503", thread=0, run=0
    04:16:46.526 r34 [WRN] http_client.go:1104> Non ok http code, code=503,
status="HTTP/1.1 503", thread=0, run=0
    04:16:46.529 r34 [WRN] http_client.go:1104> Non ok http code, code=503,
status="HTTP/1.1 503", thread=0, run=0
    04:16:46.533 r36 [WRN] http_client.go:1104> Non ok http code, code=503,
status="HTTP/1.1 503", thread=2, run=0
    04:16:46.535 r36 [WRN] http_client.go:1104> Non ok http code, code=503,
status="HTTP/1.1 503", thread=2, run=0
    Ended after 63.600371ms : 30 calls. qps=471.7
    Aggregated Function Time : count 30 avg 0.0053485212 +/- 0.004501 min 0.000676021 max
0.019712642 sum 0.160455637
    # range, mid point, percentile, count
    >= 0.000676021 <= 0.001 , 0.00083801 , 3.33, 1
    > 0.001 <= 0.002 , 0.0015 , 30.00, 8
    > 0.002 <= 0.003 , 0.0025 , 50.00, 6
    > 0.003 <= 0.004 , 0.0035 , 56.67, 2
    > 0.004 <= 0.005 , 0.0045 , 60.00, 1
    > 0.005 <= 0.006 , 0.0055 , 63.33, 1
    > 0.006 <= 0.007 , 0.0065 , 66.67, 1
    > 0.008 <= 0.009 , 0.0085 , 73.33, 2
    > 0.009 <= 0.01 , 0.0095 , 86.67, 4
    > 0.01 <= 0.011 , 0.0105 , 90.00, 1
    > 0.011 <= 0.012 , 0.0115 , 93.33, 1
    > 0.012 <= 0.014 , 0.013 , 96.67, 1
    > 0.018 <= 0.0197126 , 0.0188563 , 100.00, 1
```

```
# target 50% 0.003
# target 75% 0.009125
# target 90% 0.011
# target 99% 0.0191988
# target 99.9% 0.0196613
Error cases : count 18 avg 0.0027506623 +/- 0.002253 min 0.000676021 max 0.010483531 sum 0.049511922
# range, mid point, percentile, count
>= 0.000676021 <= 0.001 , 0.00083801 , 5.56, 1
> 0.001 <= 0.002 , 0.0015 , 50.00, 8
> 0.002 <= 0.003 , 0.0025 , 77.78, 5
> 0.003 <= 0.004 , 0.0035 , 83.33, 1
> 0.004 <= 0.005 , 0.0045 , 88.89, 1
> 0.005 <= 0.006 , 0.0055 , 94.44, 1
> 0.01 <= 0.0104835 , 0.0102418 , 100.00, 1
# target 50% 0.002
# target 75% 0.0029
# target 90% 0.0052
# target 99% 0.0103965
# target 99.9% 0.0104748
# Socket and IP used for each connection:
[0]   7 socket used, resolved to 10.107.190.211:8000, connection timing : count 7 avg 0.000173588 +/- 6.432e-05 min 0.00011326 max 0.000300204 sum 0.001215116
[1]  10 socket used, resolved to 10.107.190.211:8000, connection timing : count 10 avg 0.0002092242 +/- 0.0001633 min 7.4225e-05 max 0.00064613 sum 0.002092242
[2]   4 socket used, resolved to 10.107.190.211:8000, connection timing : count 4 avg 0.00015824225 +/- 2.654e-05 min 0.000131327 max 0.000202122 sum 0.000632969
Connection time (s) : count 21 avg 0.00018763462 +/- 0.0001211 min 7.4225e-05 max 0.00064613 sum 0.003940327
Sockets used: 21 (for perfect keepalive, would be 3)
Uniform: false, Jitter: false, Catchup allowed: true
IP addresses distribution:
10.107.190.211:8000: 21
Code 200 : 12 (40.0 %)
Code 503 : 18 (60.0 %)
Response Header Sizes : count 30 avg 92.033333 +/- 112.7 min 0 max 231 sum 2761
Response Body/Total Sizes : count 30 avg 473.43333 +/- 284.7 min 241 max 823 sum 14203
All done 30 calls (plus 0 warmup) 5.349 ms avg, 471.7 qps
```

09 将开始看到预期的熔断行为,只有40.0%的请求成功,其余的均被熔断器拦截:

```
Code 200 : 12 (40.0 %)
Code 503 : 18 (60.0 %)
```

10 查询istio-proxy状态以了解更多熔断详情,可以看到upstream_rq_pending_overflow值为25,这意味着到目前为止已经有25个调用被标记为熔断:

```
[root@k8s-master ~]# kubectl exec $FORTIO_POD -n fusing -c istio-proxy -- pilot-agent request GET stats | grep httpbin | grep pending
# 输出信息
```

```
        cluster.outbound|8000||httpbin.fusing.svc.cluster.local.circuit_breakers.default.
remaining_pending: 1
        cluster.outbound|8000||httpbin.fusing.svc.cluster.local.circuit_breakers.default.
rq_pending_open: 0
        cluster.outbound|8000||httpbin.fusing.svc.cluster.local.circuit_breakers.high.rq_p
ending_open: 0
        cluster.outbound|8000||httpbin.fusing.svc.cluster.local.upstream_rq_pending_active:
0
        cluster.outbound|8000||httpbin.fusing.svc.cluster.local.upstream_rq_pending_failur
e_eject: 0
        cluster.outbound|8000||httpbin.fusing.svc.cluster.local.upstream_rq_pending_overfl
ow: 25
        cluster.outbound|8000||httpbin.fusing.svc.cluster.local.upstream_rq_pending_total:
26
```

15.7.9 故障注入

故障注入是一种评估系统可靠性的有效方法。通过在系统中注入故障，开发者可以模拟出各种异常场景，观察系统在面对这些故障时的表现，从而发现潜在的问题并进行优化。Istio作为一个服务网格平台，提供了故障注入功能，允许开发者在不中断现有服务的情况下，进行故障模拟和测试。

故障注入类型：

- 延迟（Delay）：模拟网络延迟或系统处理延迟，以测试系统在高负载或网络不稳定时的表现。
- 中止（Abort）：模拟系统崩溃或服务内部错误，直接返回HTTP错误代码或TCP连接失败，以测试系统的容错能力和错误恢复机制。
- 重试（Retry）：虽然重试不是直接的故障注入方式，但Istio提供了灵活的重试配置，允许开发者在请求失败时自动重试，以提升服务的可用性和稳定性。

故障注入通常应用于以下场景。

- 性能测试：通过模拟高负载或网络延迟，测试系统的处理能力和响应时间。
- 容错能力测试：通过模拟系统崩溃或服务内部错误，测试系统的容错能力和错误恢复机制。
- 压力测试：在极端条件下测试系统的稳定性和可靠性，以确保系统在面对意外情况时能够正常运行。

注意事项：

- 非生产环境测试：建议在非生产环境中进行故障注入测试，以避免对真实用户造成影响。
- 合理配置：根据实际业务需求进行故障注入配置，避免盲目配置导致资源浪费或性能下降。
- 定期测试：定期对服务进行故障注入测试，以检验服务的稳定性和容错能力，及时发现并解决问题。

接下来介绍如何实现Istio的故障注入功能。

01 创建命名空间：

```
[root@k8s-master ~]# kubectl create ns breakdown
namespace/breakdown created
```

02 设置Sidecar自动注入：

```
[root@k8s-master ~]# kubectl label namespace breakdown istio-injection=enabled
namespace/breakdown labeled
```

03 创建资源清单：

```yaml
[root@k8s-master ~]# vim breakdown.yaml
apiVersion: networking.istio.io/v1beta1
kind: Gateway
metadata:
  name: breakdown-gateway
  namespace: breakdown
spec:
  selector:
    app: istio-ingressgateway
  servers:
    - port:
        number: 80
        name: http
        protocol: HTTP
      hosts:
        - "breakdown.deanit.cn"
---
# 版本1的服务
apiVersion: apps/v1
kind: Deployment
metadata:
  labels:
    app: breakdown
  name: breakdown-v1-deployment
  namespace: breakdown
spec:
  replicas: 2
  selector:
    matchLabels:
      app: breakdown
      version: v1
  template:
    metadata:
      labels:
        app: breakdown
        version: v1
    spec:
```

```yaml
      containers:
        - image: registry.cn-hangzhou.aliyuncs.com/deanmr/dean:myappv1
          name: breakdown-deployment
---
# svc，指向两个版本的服务
apiVersion: v1
kind: Service
metadata:
  name: breakdown-service
  namespace: breakdown
spec:
  ports:
    - name: http
      port: 80
      protocol: TCP
      targetPort: 80
  selector:
    app: breakdown
  type: ClusterIP
---
# 虚拟服务
apiVersion: networking.istio.io/v1beta1
kind: VirtualService
metadata:
  name: breakdown-vs
  namespace: breakdown
spec:
  hosts:
    - "*"
  gateways:
    - breakdown-gateway
  # HTTP七层路由
  http:
  - fault:
      delay:
        percentage:
          value: 100.0 # 100%的请求都会被延迟
        fixedDelay: 5s # 延迟5秒

    # 定义了一个中止的条件，当某个进度或比例达到10.0%时，以及当HTTP状态码为500时，触发中止操作
    # abort:
    #   percentage:
    #     value: 10.0
    #   httpStatus: 500
    route:
    - destination:
        host: breakdown-service
        port:
          number: 80
```

04 应用资源：

```
[root@k8s-master ~]# kubectl apply -f breakdown.yaml
gateway.networking.istio.io/breakdown-gateway created
deployment.apps/breakdown-v1-deployment created
service/breakdown-service created
virtualservice.networking.istio.io/breakdown-vs created
```

05 查看资源状态信息：

```
[root@k8s-master ~]# kubectl get pods,svc -n breakdown
NAME                                                READY   STATUS    RESTARTS   AGE
pod/breakdown-v1-deployment-755cf68744-tcsql        2/2     Running   0          16m
pod/breakdown-v1-deployment-755cf68744-x2wsn        2/2     Running   0          16m

NAME                        TYPE        CLUSTER-IP       EXTERNAL-IP   PORT(S)   AGE
service/breakdown-service   ClusterIP   10.100.254.28    <none>        80/TCP    16m

NAME                                                  GATEWAYS                HOSTS     AGE
virtualservice.networking.istio.io/breakdown-vs       ["breakdown-gateway"]   ["*"]     16m
```

06 在k8s-master节点配置hosts进行域名解析，将域名解析到istio-ingressgateway的LoadBalancer分配的IP地址上：

```
[root@k8s-master ~]# echo "192.168.1.21 breakdown.deanit.cn">>/etc/hosts
```

07 进行请求测试，访问域名测试效果，通过加入time命令参数可以看到请求时间为5秒：

```
[root@k8s-master ~]# time curl http://breakdown.deanit.cn
Hello MyApp | Version: v1 | <a href="hostname.html">Pod Name</a>

real    0m5.023s
user    0m0.001s
sys     0m0.005s
```

至此，我们成功设置了Istio的故障注入功能。

15.8 本章小结

本章介绍了如何在Kubernetes环境中安装和使用Istio服务网格。首先，我们详细阐述了Istio的基本概念、主要组件，以及如何在K8s集群上部署Istio，从下载Istio到使用Istioctl工具完成安装，确保每一步都清晰易懂。安装完成后，通过一系列精心设计的实例，让读者亲身体验Istio流量治理的强大功能。通过本章的学习，读者不仅能够掌握在K8s上安装Istio的基本技能，还能通过实践加深对Istio服务网格各项功能的理解。

第 16 章

Kubernetes服务部署实战

本章将详细介绍如何在Kubernetes中部署各种类型的服务。从简单的单实例应用到复杂的多副本、自动扩展的分布式应用，通过这些应用，读者可以掌握如何使用YAML配置文件定义资源对象，包括Pods、Deployments、Services等，并通过Kubectl命令行工具或Kubernetes Dashboard进行部署和管理。

16.1　K8s部署MinIO开源对象存储

MinIO是一个高性能的分布式对象存储服务，它兼容Amazon S3云存储服务接口。通过在Kubernetes集群中部署MinIO，可以利用Kubernetes的自动化部署、扩展和管理能力来运行一个高可用、可扩展的MinIO存储系统。

下面介绍部署MinIO的具体操作步骤。

01 创建命名空间：

```
[root@k8s-master ~]# kubectl create ns minio
namespace/minio created
```

02 创建资源清单：

```
[root@k8s-master ~]# vim minio.yaml
---
# pvc动态申请PV
apiVersion: v1
kind: PersistentVolumeClaim
metadata:
  name: minio-storage
  namespace: minio
spec:
  accessModes:
    - ReadWriteMany
```

```yaml
      storageClassName: dean-nfs
      resources:
        requests:
          storage: 5Gi
---
# pvc动态申请PV
apiVersion: v1
kind: PersistentVolumeClaim
metadata:
  name: minio-config
  namespace: minio
spec:
  accessModes:
    - ReadWriteMany
  storageClassName: dean-nfs
  resources:
    requests:
      storage: 1Gi
---
# 创建控制器
apiVersion: apps/v1
kind: Deployment
metadata:
  namespace: minio
  name: minio
  labels:
    component: minio
spec:
  strategy:
    type: Recreate
  selector:
    matchLabels:
      component: minio
  template:
    metadata:
      labels:
        component: minio
    spec:
      containers:
        - name: minio
          image: docker.m.daocloud.io/minio/minio:latest
          imagePullPolicy: IfNotPresent
          args:
          - server
          - /storage
          - --config-dir=/config
          - --console-address=:9001
          env:
          # minio的用户
          - name: MINIO_ACCESS_KEY
```

```yaml
          value: "dean"
        # minio的密码
        - name: MINIO_SECRET_KEY
          value: "deanpswd"
        ports:
        - containerPort: 9000
        - containerPort: 9001
        volumeMounts:
        - name: storage
          mountPath: "/storage"
        - name: config
          mountPath: "/config"
      volumes:
        - name: storage
          persistentVolumeClaim:
            claimName: minio-storage
        - name: config
          persistentVolumeClaim:
            claimName: minio-config
---
# 创建SVC
apiVersion: v1
kind: Service
metadata:
  namespace: minio
  name: minio
  labels:
    component: minio
spec:
  type: ClusterIP
  ports:
    - name: api
      port: 9000
      targetPort: 9000
    - name: console
      port: 9001
      targetPort: 9001
  selector:
    component: minio
---
apiVersion: networking.k8s.io/v1
kind: Ingress
metadata:
  name: minio
  namespace: minio
spec:
  ingressClassName: nginx
  rules:
  - host: minio.deanit.cn
    http:
```

```yaml
      paths:
      - backend:
          service:
            name: minio
            port:
              number: 9001
        path: /
        pathType: Prefix
```

03 应用资源：

```
[root@k8s-master ~]# kubectl apply -f minio.yaml
persistentvolumeclaim/minio-storage created
persistentvolumeclaim/minio-config created
deployment.apps/minio created
service/minio created
ingress.networking.k8s.io/minio created
```

04 查看Minio服务各资源的状态，如图16-1所示。

图 16-1　Minio 服务各资源的状态

05 由于我们创建了Ingress资源，因此可以通过域名来访问Minio服务，各客户端配置如下：

```
# Linux和macOS客户端配置
$ vim /etc/hosts
192.168.1.11 minio.deanit.cn


# Windows客户端配置
# hosts文件位于路径：C:\Windows\System32\drivers\etc\hosts
192.168.1.11 minio.deanit.cn
```

> 注意　由于hosts文件属于系统文件，因此在修改之前需要获取管理员权限。

06 通过域名访问后，Minio登录页面如图16-2所示。

07 输入资源清单中设置的账号和密码，账号为dean，密码为deanpswd。登录成功后，Minio页面如图16-3所示。

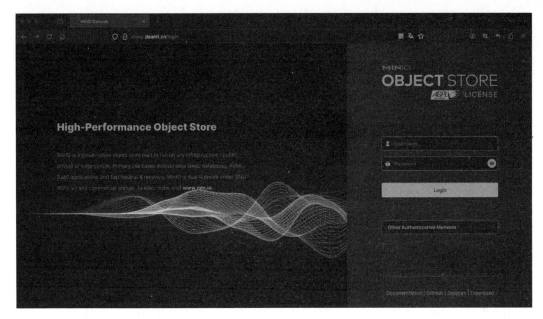

图 16-2　Minio 登录页面

图 16-3　Minio 登录成功页面

16.2　K8s部署Metabase数据库连接工具

Metabase数据库连接工具是一款开源的商业智能（Business Intelligence，BI）和数据可视化工具,以其易用性、灵活性、可扩展性和开源性而受到广泛欢迎。Metabase支持连接多种数据源,包括MySQL、PostgreSQL、Oracle、SQL Server等常见的关系数据库,以及NoSQL数据库,如MongoDB,还有云服务如Amazon Redshift、Google BigQuery等。这使得用户可以根据自己的需求方便地连接到不同的数据源。Metabase提供了多种数据可视化方式,包括柱状图、折线图、饼图、散点图等,用户可以根据自己的需求选择适合的图表类型。此外,Metabase还支持图表的自定义,用户可以调整图表的颜色、字体、标签等,使报表更加美观和易读。

下面介绍在Kubernetes集群中部署Metabase工具的操作步骤。

01 创建命名空间:

```
[root@k8s-master ~]# kubectl create ns ops
namespace/ops created
```

02 创建资源清单:

```
[root@k8s-master ~]# vim metabase.yaml
---
apiVersion: apps/v1
kind: StatefulSet

metadata:
  name: metabase
  namespace: ops

spec:
  serviceName: metabase
  selector:
    matchLabels:
      name: metabase
  template:
    metadata:
      labels:
        name: metabase
    spec:
      containers:
      - name: metabase
        image: metabase/metabase:latest
        ports:
        - name: http
          protocol: TCP
          containerPort: 3000
        volumeMounts:
        - name: datadir
          mountPath: /metabase.db
  volumeClaimTemplates:
  - metadata:
      name: datadir
```

```yaml
  spec:
    accessModes: ["ReadWriteOnce"]
    storageClassName: dean-nfs
    resources:
      requests:
        storage: 1Gi
---
apiVersion: v1
kind: Service

metadata:
  name: metabase
  namespace: ops
spec:
  type: ClusterIP
  selector:
    name: metabase
  ports:
  - name: http
    port: 3000
    targetPort: 3000
---
apiVersion: networking.k8s.io/v1
kind: Ingress

metadata:
  name: metabase
  namespace: ops

spec:
  ingressClassName: nginx
  rules:
  - host: metabase.deanit.cn
    http:
      paths:
      - pathType: Prefix
        path: /
        backend:
          service:
            name: metabase
            port:
              number: 3000
```

03 应用资源：

```
[root@k8s-master ~]# kubectl apply -f metabase.yaml
statefulset.apps/metabase created
service/metabase created
ingress.networking.k8s.io/metabase created
```

04 查看Metabase服务各资源的状态，如图16-4所示。

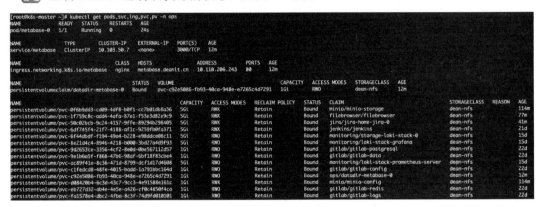

图 16-4　Metabase 服务各资源的状态

05 由于我们创建了Ingress资源，因此可以通过域名来访问Metabase服务，各客户端配置如下：

```
# Linux和macOS客户端配置
$ vim /etc/hosts
192.168.1.11 metabase.deanit.cn


# Windows客户端配置
# hosts文件位于路径：C:\Windows\System32\drivers\etc\hosts。
192.168.1.11 metabase.deanit.cn
```

> **注意**　由于hosts文件属于系统文件，因此在修改之前需要获取管理员权限。

06 通过域名访问后，Metabase初始化开始页面如图16-5所示。

图 16-5　Metabase 初始化开始页面

07 单击"让我们开始吧"按钮，依次进行配置，初始化完成后，页面显示如图16-6所示。

图 16-6　Metabase 初始化完成

08 单击Sample Dataset数据库，选择People表，如图16-7所示。

图 16-7　示例数据库表信息

09 单击右上角的 ⚙ 按钮，依次单击"Metabase管理员"→"添加数据库"，可以看到支持现在主流数据库的选项，如图16-8所示。

图 16-8　配置数据库信息

16.3　K8s部署phpMyAdmin数据库连接工具

phpMyAdmin是一个流行的开源工具，可以通过Web界面管理MySQL和MariaDB数据库。它提供了一个图形界面，让用户可以执行各种数据库管理任务，而无须编写复杂的SQL语句或使用命令行工具。phpMyAdmin支持大多数MySQL功能，包括但不限于创建和删除数据库、管理数据库表、编辑表结构、运行SQL查询、管理用户权限等。

下面介绍在Kubernetes集群中部署phpMyAdmin工具的操作步骤。

01 创建命名空间：

```
[root@k8s-master ~]# kubectl create ns ops
namespace/ops created
```

02 创建资源清单:

```
[root@k8s-master ~]# vim myadmin.yaml
---
apiVersion: apps/v1
kind: Deployment
metadata:
  name: myadmin
  namespace: ops
  labels:
    app: myadmin
spec:
  replicas: 1
  selector:
    matchLabels:
      app: myadmin
  template:
    metadata:
      labels:
        app: myadmin
    spec:
      containers:
      - image: phpmyadmin:latest
        imagePullPolicy: IfNotPresent
        name: phpmyadmin
        ports:
        - containerPort: 80
          name: http
          protocol: TCP
        env:
          # 允许连接多个 MySQL 数据库
        - name: PMA_ARBITRARY
          value: '1'
---
apiVersion: v1
kind: Service
metadata:
  name: myadmin
  namespace: ops
  labels:
    app: myadmin
spec:
  type: ClusterIP
  ports:
  - name: http
    port: 80
    protocol: TCP
    targetPort: http
  selector:
    app: myadmin
```

```yaml
---
apiVersion: networking.k8s.io/v1
kind: Ingress

metadata:
  name: myadmin
  namespace: ops

spec:
  ingressClassName: nginx
  rules:
  - host: myadmin.deanit.cn
    http:
      paths:
      - pathType: Prefix
        path: /
        backend:
          service:
            name: myadmin
            port:
              number: 80
```

03 应用资源：

```
[root@k8s-master ~]# kubectl apply -f myadmin.yaml
deployment.apps/myadmin created
service/myadmin created
ingress.networking.k8s.io/myadmin created
```

04 查看Myadmin服务各资源的状态，如图16-9所示。

图 16-9　Myadmin 服务各资源的状态

05 由于我们创建了Ingress资源，因此可以通过域名来访问Metabase服务，各客户端配置如下：

```
# Linux和macOS客户端配置
$ vim /etc/hosts
192.168.1.11 myadmin.deanit.cn


# Windows客户端配置
# hosts文件位于路径：C:\Windows\System32\drivers\etc\hosts。
192.168.1.11 myadmin.deanit.cn
```

> **注意** 由于hosts文件属于系统文件，因此在修改之前需要获取管理员权限。

06 通过域名访问，输入MySQL数据库的IP:Port、用户名及密码，登录即可，如图16-10所示。

图 16-10　Myadmin 服务登录页面

07 登录成功后如图16-11所示。

图 16-11　Myadmin 服务登录成功

16.4 K8s部署Nacos配置中心

Nacos是阿里巴巴开源的一款用于动态服务发现、配置管理和服务管理的平台，全称为Dynamic Naming and Configuration Service。Nacos旨在帮助开发人员构建云原生应用，通过提供一系列简单易用的特性集，快速实现微服务架构中的服务注册与发现、服务配置管理、服务元数据及流量管理等关键功能。

Nacos核心功能包括以下几点。

1. 服务注册与发现

- 应用程序启动后，可以通过Nacos客户端将服务实例信息（如服务名、IP 地址、端口号等）注册到Nacos服务器上。
- 其他服务可以通过Nacos查询注册中心，获取所需服务的实例列表，并根据负载均衡策略调用服务。
- Nacos支持多种注册方式（如HTTP、DNS等），并能动态地管理服务实例信息，实现服务之间的高效通信。

2. 配置管理

- Nacos提供了一个集中的配置管理系统，支持动态配置更新、配置监听和配置推送等功能。
- 开发人员可以通过Nacos管理各种配置信息，当配置发生变化时，Nacos会自动推送到应用中，确保配置的一致性和实时性。
- Nacos支持多环境配置管理，方便在不同环境（如开发、测试、生产）中管理配置。

3. 服务健康检查

- Nacos会定期检查已注册的服务实例的健康状态，对于不健康的实例进行标记，并在服务发现时过滤掉，确保服务的可靠性和稳定性。
- 通过心跳机制和健康检查，Nacos能够及时发现并处理不健康的服务实例。

下面介绍在Kubernetes集群中部署Nacos的操作步骤。

01 创建命名空间：

```
[root@k8s-master ~]# kubectl create ns nacos
namespace/nacos created
```

02 创建资源清单：

```
[root@k8s-master ~]# vim nacos.yaml
---
apiVersion: apps/v1
kind: Deployment
```

```yaml
metadata:
  name: nacos
  namespace: nacos
spec:
  replicas: 1
  selector:
    matchLabels:
      app: nacos
  template:
    metadata:
      labels:
        app: nacos
    spec:
      containers:
      - name: nacos
        image: docker.m.daocloud.io/nacos/nacos-server:v2.2.0
        env:
        - name: SPRING_DATASOURCE_PLATFORM
          value: "mysql"
        - name: MYSQL_SERVICE_HOST
          value: "192.168.1.254"
        - name: MYSQL_SERVICE_DB_NAME
          value: "nacos"
        - name: MYSQL_SERVICE_PORT
          value: "3306"
        - name: MYSQL_SERVICE_USER
          value: "root"
        - name: MYSQL_SERVICE_PASSWORD
          value: "deanpswd"
        - name: PREFER_HOST_MODE
          value: "hostname"
        - name: MODE
          value: "standalone"
        ports:
        - containerPort: 8848
---
apiVersion: v1
kind: Service
metadata:
  name: nacos
  namespace: nacos
spec:
  type: NodePort
  ports:
    - port: 8848
      name: port-8848
      targetPort: 8848
      nodePort: 30848
    - port: 9849
      name: port-9849
```

```yaml
      targetPort: 9849
    - port: 9848
      name: port-9848
      targetPort: 9848
  selector:
    app: nacos
---
apiVersion: v1
kind: Service
metadata:
  name: nacos-headless
  namespace: nacos
  labels:
    app: nacos
spec:
  ports:
    - port: 8848
      name: server
      targetPort: 8848
    - port: 9848
      name: server-1
      targetPort: 9848
    - port: 9849
      name: server-2
      targetPort: 9849
    - port: 7848
      name: rpc
      targetPort: 7848
  clusterIP: None
  selector:
    app: nacos
```

03 通过数据库连接工具，连接到我们的MySQL数据库，新建名为nacos的库，如图16-12所示。

图16-12　新建数据库

04 在新建的名为nacos的库中执行以下初始化SQL脚本：

```sql
/******************************************/
/*   表名称 = config_info   */
/******************************************/
CREATE TABLE `config_info` (
  `id` bigint(20) NOT NULL AUTO_INCREMENT COMMENT 'id',
  `data_id` varchar(255) NOT NULL COMMENT 'data_id',
  `group_id` varchar(255) DEFAULT NULL,
  `content` longtext NOT NULL COMMENT 'content',
  `md5` varchar(32) DEFAULT NULL COMMENT 'md5',
  `gmt_create` datetime NOT NULL DEFAULT CURRENT_TIMESTAMP COMMENT '创建时间',
  `gmt_modified` datetime NOT NULL DEFAULT CURRENT_TIMESTAMP COMMENT '修改时间',
  `src_user` text COMMENT 'source user',
  `src_ip` varchar(50) DEFAULT NULL COMMENT 'source ip',
  `app_name` varchar(128) DEFAULT NULL,
  `tenant_id` varchar(128) DEFAULT '' COMMENT '租户字段',
  `c_desc` varchar(256) DEFAULT NULL,
  `c_use` varchar(64) DEFAULT NULL,
  `effect` varchar(64) DEFAULT NULL,
  `type` varchar(64) DEFAULT NULL,
  `c_schema` text,
  `encrypted_data_key` text NOT NULL COMMENT '秘钥',
  PRIMARY KEY (`id`),
  UNIQUE KEY `uk_configinfo_datagrouptenant` (`data_id`,`group_id`,`tenant_id`)
) ENGINE=InnoDB DEFAULT CHARSET=utf8 COLLATE=utf8_bin COMMENT='config_info';

/******************************************/
/*   表名称 = config_info_aggr   */
/******************************************/
CREATE TABLE `config_info_aggr` (
  `id` bigint(20) NOT NULL AUTO_INCREMENT COMMENT 'id',
  `data_id` varchar(255) NOT NULL COMMENT 'data_id',
  `group_id` varchar(255) NOT NULL COMMENT 'group_id',
  `datum_id` varchar(255) NOT NULL COMMENT 'datum_id',
  `content` longtext NOT NULL COMMENT '内容',
  `gmt_modified` datetime NOT NULL COMMENT '修改时间',
  `app_name` varchar(128) DEFAULT NULL,
  `tenant_id` varchar(128) DEFAULT '' COMMENT '租户字段',
  PRIMARY KEY (`id`),
  UNIQUE KEY `uk_configinfoaggr_datagrouptenantdatum` (`data_id`,`group_id`,`tenant_id`,`datum_id`)
) ENGINE=InnoDB DEFAULT CHARSET=utf8 COLLATE=utf8_bin COMMENT='增加租户字段';

/******************************************/
/*   表名称 = config_info_beta   */
/******************************************/
CREATE TABLE `config_info_beta` (
  `id` bigint(20) NOT NULL AUTO_INCREMENT COMMENT 'id',
  `data_id` varchar(255) NOT NULL COMMENT 'data_id',
  `group_id` varchar(128) NOT NULL COMMENT 'group_id',
  `app_name` varchar(128) DEFAULT NULL COMMENT 'app_name',
```

```sql
  `content` longtext NOT NULL COMMENT 'content',
  `beta_ips` varchar(1024) DEFAULT NULL COMMENT 'betaIps',
  `md5` varchar(32) DEFAULT NULL COMMENT 'md5',
  `gmt_create` datetime NOT NULL DEFAULT CURRENT_TIMESTAMP COMMENT '创建时间',
  `gmt_modified` datetime NOT NULL DEFAULT CURRENT_TIMESTAMP COMMENT '修改时间',
  `src_user` text COMMENT 'source user',
  `src_ip` varchar(50) DEFAULT NULL COMMENT 'source ip',
  `tenant_id` varchar(128) DEFAULT '' COMMENT '租户字段',
  `encrypted_data_key` text NOT NULL COMMENT '秘钥',
  PRIMARY KEY (`id`),
  UNIQUE KEY `uk_configinfobeta_datagrouptenant` (`data_id`,`group_id`,`tenant_id`)
) ENGINE=InnoDB DEFAULT CHARSET=utf8 COLLATE=utf8_bin COMMENT='config_info_beta';

/******************************************/
/*   表名称 = config_info_tag   */
/******************************************/
CREATE TABLE `config_info_tag` (
  `id` bigint(20) NOT NULL AUTO_INCREMENT COMMENT 'id',
  `data_id` varchar(255) NOT NULL COMMENT 'data_id',
  `group_id` varchar(128) NOT NULL COMMENT 'group_id',
  `tenant_id` varchar(128) DEFAULT '' COMMENT 'tenant_id',
  `tag_id` varchar(128) NOT NULL COMMENT 'tag_id',
  `app_name` varchar(128) DEFAULT NULL COMMENT 'app_name',
  `content` longtext NOT NULL COMMENT 'content',
  `md5` varchar(32) DEFAULT NULL COMMENT 'md5',
  `gmt_create` datetime NOT NULL DEFAULT CURRENT_TIMESTAMP COMMENT '创建时间',
  `gmt_modified` datetime NOT NULL DEFAULT CURRENT_TIMESTAMP COMMENT '修改时间',
  `src_user` text COMMENT 'source user',
  `src_ip` varchar(50) DEFAULT NULL COMMENT 'source ip',
  PRIMARY KEY (`id`),
  UNIQUE KEY `uk_configinfotag_datagrouptenanttag` (`data_id`,`group_id`,`tenant_id`,`tag_id`)
) ENGINE=InnoDB DEFAULT CHARSET=utf8 COLLATE=utf8_bin COMMENT='config_info_tag';

/******************************************/
/*   表名称 = config_tags_relation   */
/******************************************/
CREATE TABLE `config_tags_relation` (
  `id` bigint(20) NOT NULL COMMENT 'id',
  `tag_name` varchar(128) NOT NULL COMMENT 'tag_name',
  `tag_type` varchar(64) DEFAULT NULL COMMENT 'tag_type',
  `data_id` varchar(255) NOT NULL COMMENT 'data_id',
  `group_id` varchar(128) NOT NULL COMMENT 'group_id',
  `tenant_id` varchar(128) DEFAULT '' COMMENT 'tenant_id',
  `nid` bigint(20) NOT NULL AUTO_INCREMENT,
  PRIMARY KEY (`nid`),
  UNIQUE KEY `uk_configtagrelation_configidtag` (`id`,`tag_name`,`tag_type`),
  KEY `idx_tenant_id` (`tenant_id`)
) ENGINE=InnoDB DEFAULT CHARSET=utf8 COLLATE=utf8_bin COMMENT='config_tag_relation';

/******************************************/
/*   表名称 = group_capacity   */
```

```sql
/*********************************************/
CREATE TABLE `group_capacity` (
  `id` bigint(20) unsigned NOT NULL AUTO_INCREMENT COMMENT '主键ID',
  `group_id` varchar(128) NOT NULL DEFAULT '' COMMENT 'Group ID,空字符表示整个集群',
  `quota` int(10) unsigned NOT NULL DEFAULT '0' COMMENT '配额,0表示使用默认值',
  `usage` int(10) unsigned NOT NULL DEFAULT '0' COMMENT '使用量',
  `max_size` int(10) unsigned NOT NULL DEFAULT '0' COMMENT '单个配置大小上限,单位为字节,0表示使用默认值',
  `max_aggr_count` int(10) unsigned NOT NULL DEFAULT '0' COMMENT '聚合子配置最大个数,0表示使用默认值',
  `max_aggr_size` int(10) unsigned NOT NULL DEFAULT '0' COMMENT '单个聚合数据的子配置大小上限,单位为字节,0表示使用默认值',
  `max_history_count` int(10) unsigned NOT NULL DEFAULT '0' COMMENT '最大变更历史数量',
  `gmt_create` datetime NOT NULL DEFAULT CURRENT_TIMESTAMP COMMENT '创建时间',
  `gmt_modified` datetime NOT NULL DEFAULT CURRENT_TIMESTAMP COMMENT '修改时间',
  PRIMARY KEY (`id`),
  UNIQUE KEY `uk_group_id` (`group_id`)
) ENGINE=InnoDB DEFAULT CHARSET=utf8 COLLATE=utf8_bin COMMENT='集群、各Group容量信息表';

/*********************************************/
/*      表名称 = his_config_info              */
/*********************************************/
CREATE TABLE `his_config_info` (
  `id` bigint(20) unsigned NOT NULL,
  `nid` bigint(20) unsigned NOT NULL AUTO_INCREMENT,
  `data_id` varchar(255) NOT NULL,
  `group_id` varchar(128) NOT NULL,
  `app_name` varchar(128) DEFAULT NULL COMMENT 'app_name',
  `content` longtext NOT NULL,
  `md5` varchar(32) DEFAULT NULL,
  `gmt_create` datetime NOT NULL DEFAULT CURRENT_TIMESTAMP,
  `gmt_modified` datetime NOT NULL DEFAULT CURRENT_TIMESTAMP,
  `src_user` text,
  `src_ip` varchar(50) DEFAULT NULL,
  `op_type` char(10) DEFAULT NULL,
  `tenant_id` varchar(128) DEFAULT '' COMMENT '租户字段',
  `encrypted_data_key` text NOT NULL COMMENT '秘钥',
  PRIMARY KEY (`nid`),
  KEY `idx_gmt_create` (`gmt_create`),
  KEY `idx_gmt_modified` (`gmt_modified`),
  KEY `idx_did` (`data_id`)
) ENGINE=InnoDB DEFAULT CHARSET=utf8 COLLATE=utf8_bin COMMENT='多租户改造';

/*********************************************/
/*      表名称 = tenant_capacity              */
/*********************************************/
CREATE TABLE `tenant_capacity` (
  `id` bigint(20) unsigned NOT NULL AUTO_INCREMENT COMMENT '主键ID',
  `tenant_id` varchar(128) NOT NULL DEFAULT '' COMMENT 'Tenant ID',
  `quota` int(10) unsigned NOT NULL DEFAULT '0' COMMENT '配额,0表示使用默认值',
```

```sql
  `usage` int(10) unsigned NOT NULL DEFAULT '0' COMMENT '使用量',
  `max_size` int(10) unsigned NOT NULL DEFAULT '0' COMMENT '单个配置大小上限,单位为字节,0表示使用默认值',
  `max_aggr_count` int(10) unsigned NOT NULL DEFAULT '0' COMMENT '聚合子配置最大个数',
  `max_aggr_size` int(10) unsigned NOT NULL DEFAULT '0' COMMENT '单个聚合数据的子配置大小上限,单位为字节,0表示使用默认值',
  `max_history_count` int(10) unsigned NOT NULL DEFAULT '0' COMMENT '最大变更历史数量',
  `gmt_create` datetime NOT NULL DEFAULT CURRENT_TIMESTAMP COMMENT '创建时间',
  `gmt_modified` datetime NOT NULL DEFAULT CURRENT_TIMESTAMP COMMENT '修改时间',
  PRIMARY KEY (`id`),
  UNIQUE KEY `uk_tenant_id` (`tenant_id`)
) ENGINE=InnoDB DEFAULT CHARSET=utf8 COLLATE=utf8_bin COMMENT='租户容量信息表';

CREATE TABLE `tenant_info` (
  `id` bigint(20) NOT NULL AUTO_INCREMENT COMMENT 'id',
  `kp` varchar(128) NOT NULL COMMENT 'kp',
  `tenant_id` varchar(128) default '' COMMENT 'tenant_id',
  `tenant_name` varchar(128) default '' COMMENT 'tenant_name',
  `tenant_desc` varchar(256) DEFAULT NULL COMMENT 'tenant_desc',
  `create_source` varchar(32) DEFAULT NULL COMMENT 'create_source',
  `gmt_create` bigint(20) NOT NULL COMMENT '创建时间',
  `gmt_modified` bigint(20) NOT NULL COMMENT '修改时间',
  PRIMARY KEY (`id`),
  UNIQUE KEY `uk_tenant_info_kptenantid` (`kp`,`tenant_id`),
  KEY `idx_tenant_id` (`tenant_id`)
) ENGINE=InnoDB DEFAULT CHARSET=utf8 COLLATE=utf8_bin COMMENT='tenant_info';

CREATE TABLE `users` (
    `username` varchar(50) NOT NULL PRIMARY KEY,
    `password` varchar(500) NOT NULL,
    `enabled` boolean NOT NULL
);

CREATE TABLE `roles` (
    `username` varchar(50) NOT NULL,
    `role` varchar(50) NOT NULL,
    UNIQUE INDEX `idx_user_role` (`username` ASC, `role` ASC) USING BTREE
);

CREATE TABLE `permissions` (
    `role` varchar(50) NOT NULL,
    `resource` varchar(255) NOT NULL,
    `action` varchar(8) NOT NULL,
    UNIQUE INDEX `uk_role_permission` (`role`,`resource`,`action`) USING BTREE
);

INSERT INTO users (username, password, enabled) VALUES ('nacos', '$2a$10$EuWPZHzz32dJN7jexM34MOeYirDdFAZm2kuWj7VEOJhhZkDrxfvUu', TRUE);

INSERT INTO roles (username, role) VALUES ('nacos', 'ROLE_ADMIN');
```

如果打算在Nacos中使用外部数据库来存储数据,那么确实需要自己初始化数据库和表结构。

但如果只是在测试或开发环境中使用Nacos,并且接受其默认的嵌入式数据库配置,那么就不需要自己进行SQL初始化了。在本示例中,我们将使用外部数据库来存储数据。

05 应用资源:

```
[root@k8s-master ~]# kubectl apply -f nacos.yaml
deployment.apps/nacos created
service/nacos created
service/nacos-headless created
```

06 查看Nacos服务资源信息,如图16-13所示。

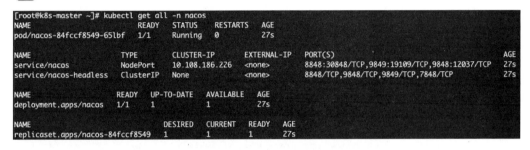

图 16-13　Nacos 服务资源信息

07 使用浏览器通过NodePort端口进行访问,如图16-14所示。Nacos通常为内部服务,不要暴露到公网上。

图 16-14　Nacos 登录页面

08 Nacos默认的账号和密码都是nacos,登录成功后,如图16-15所示。

图 16-15　Nacos 登录成功页面

09 查看集群下的节点列表信息，如图16-16所示。

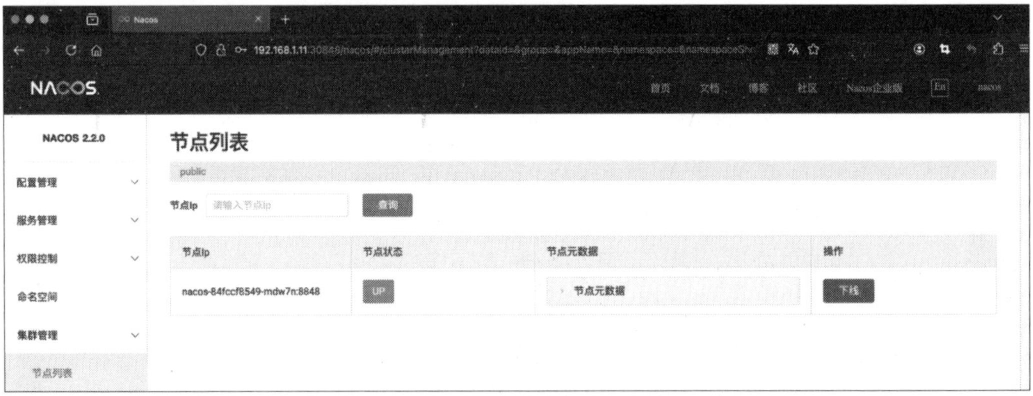

图 16-16　Nacos 集群节点列表信息

16.5　本章小结

本章介绍了在Kubernetes环境中部署多种服务的详细过程，供助实例逐一展示了如何编写相应的Kubernetes资源清单文件（如Deployments、Services、ConfigMaps、Secrets等），并通过Kubectl命令行工具将这些服务部署到K8s集群中。每个实例都详细记录了部署步骤、配置选项，旨在帮助读者理解并掌握K8s服务的安装与管理过程。通过这些实例，读者能够学习如何在K8s中部署单个服务，掌握在K8s环境中快速部署和运维各种服务的技能，从而更好地利用Kubernetes构建高效、可扩展、可靠的应用架构。